DEFENSE ENERGY MANAGEMENT

DEFENSE ENERGY MANAGEMENT

EDWARD R. MYERS
EDITOR

Nova Science Publishers, Inc.
New York

For permission to use material from this book please contact us:
Telephone 631-231-7269; Fax 631-231-8175
Web Site: http://www.novapublishers.com

LIBRARY OF CONGRESS CATALOGING-IN-PUBLICATION DATA

Available upon request.

ISBN: 978-1-60692-574-4

Published by Nova Science Publishers, Inc. ✢ New York

CONTENTS

PREFACE

The Department of Defense (DoD) continues to make significant progress toward achieving the goals of the Energy Policy Act of 2005 (EPAct 2005) and Executive Order (EO) 13423, Strengthening Federal Environmental, Energy, and Transportation Management. The DOD Energy Program initiatives include energy awareness efforts, energy manager training, audit programs, procurement of energy efficient products, and the use of sustainable design in new construction and major renovation. Other contributing factors include integrated energy planning, enhanced use of renewable energy, demonstration of innovative technologies, and the use of Energy Savings Performance Contracts (ESPC) and Utility Energy Service Contracts (UESC). DOD is responding to EPAct 2005 and EO 13423. Combined, these mandates established a new energy baseline (2003), increased the annual reduction requirement to 3 percent per year, increased the percentage of renewable energy required (7.5 percent by 2013), increased energy efficiency of new construction to 30 percent below the current standard, and required metering electricity consumption of all facilities. Through Fiscal Year (FY) 2007, the Department of Defense achieved a 10.1 percent decrease in goal facility energy consumption (as measured on a British Thermal Units (Btu) per gross square foot (GSF) basis [Btu/GSF]) as compared to the revised 2003 baseline. The Department of the Army determined that the square footage was over reported in 2003, compared to data contained in the real property database. Therefore, this chapter contains a significantly revised baseline, which raises the previously reported Btu/GSF from 113,510 to 116,134. At the end of FY 2007 the Department has 1.95 billion square feet of facilities and spent $3.4 billion on facility energy. DoD spent $9.5 billion on non-fleet vehicles and other equipment – such as auto gasoline, LPG-Propane, Aviation Gasoline, jet fuel and Navy-special fuel. DOD continues to make progress in installing renewable energy technologies and purchasing electricity generated from renewable sources (solar, wind, geothermal, and biomass) when life cycle cost-effective. The National Defense Authorization Act of 2007 codified a 2005 DOD goal to produce or procure renewable energy equivalent to 25 percent of facility electrical consumption. The total renewable energy that the Department produced or procured in FY 2007 amounted to 12,054 trillion Btu and represents 11.9 percent of the facility electrical consumption. For FY 2007, the Department of Energy revised the guidance for compliance with the renewable energy requirements of EPAct 2005 and EO 13423, allowing only renewable electricity. Under this revised guidance, DoD achieved 5.5 percent total and 3.3 percent new renewable energy, well exceeding the goals of 3 percent and 1.5 percent respectively.

Chapter 2 - Several issues, such as rising fuel costs, worldwide energy demand, and the high fuel burden during operations, underscore the importance of energy to DOD. Fuel costs for DOD are substantial and the volatility of world oil prices will likely continue to affect the department—which may require DOD to make difficult trade-offs, such as redirecting funds from ongoing programs to pay for needed fuel. In addition, both the Army and Marine Corps have plans to grow their forces over the next several years, which will inevitably require larger amounts of fuel to sustain these forces and their weapons systems. Other energy issues that are likely to affect DOD in the future are the increased U.S. dependence on foreign oil, projected increases in the worldwide demand for oil, and uncertainties about world oil supplies. Furthermore, DOD's high fuel requirements on the battlefield can place a significant logistics burden on military forces; limit the range and pace of operations; and add to mission risks, including exposing supply convoys to attack. Given these issues, DOD must be well positioned to effectively manage energy demands for military operations.

DOD and the military services have several initiatives under way to reduce demand for mobility energy. At the department level, the Office of the Secretary of Defense (OSD) created a task force in 2006 to address energy security concerns. Moreover, in 2007, the Deputy Secretary of Defense included energy in DOD's list of the top 25 transformational priorities for the department as part of its initiative to pursue targeted acquisition reforms. Each of the military services also has its own initiatives under way to reduce mobility energy demand. The Army is addressing fuel consumption at forward-deployed locations by developing foam-insulated tents and temporary dome structures that are more efficient to heat and cool, reducing the demand for fuel-powered generators. The Navy has established an energy conservation program to encourage ships to reduce energy consumption. The Air Force has developed an energy strategy and undertaken initiatives to determine fuel-efficient flight routes, reduce the weight on aircraft, optimize air refueling, and improve the efficiency of ground operations. The Marine Corps has initiated research and development efforts to develop alternative power sources, such as hybrid power, and improve fuel management.

While these and other individual efforts are under way to reduce mobility energy demand, we found that DOD does not have an overarching organizational framework to guide and oversee these efforts. Our prior work has shown that an overarching organizational framework is critical to successful transformation in both public and private organizations. Key elements of such a framework include (1) top-level leadership and an implementation team with dedicated resources and funding; (2) a comprehensive strategic plan that includes goals, objectives, methods, timelines, and outcome-oriented performance metrics; and (3) a communication strategy that provides shared expectations and reports related progress. We found that DOD's current approach to mobility energy lacks these elements. For example, while DOD has begun to increase management attention on energy issues, it has not designated a single executive-level OSD official—supported by an implementation team—who is accountable for mobility energy matters across the department, who participates in top policy-making decisions as an advocate for reducing mobility energy demand, and who serves as a stakeholder in interagency discussions about national energy concerns. Currently, DOD's approach to mobility energy is decentralized, with fuel oversight and management responsibilities diffused among several OSD and military service offices as well as working groups. In addition, until DOD fully develops and implements a comprehensive strategic plan for mobility energy, it cannot be certain that mobility energy reduction efforts align with the department's energy mission or strategic goals to ensure that they are appropriately prioritized or to know whether critical gaps or duplication of efforts exist. Finally, without an effective mechanism to facilitate communication of mobility energy reduction efforts among OSD and the military services, DOD cannot be assured that these efforts are consistent with DOD's energy priorities and goals. We also found that DOD has made limited progress in incorporating fuel efficiency as a consideration in key business processes—which include developing requirements for and acquiring new weapons systems—and in implementing recommendations from department-sponsored studies on fuel reduction. With a mobility energy overarching organizational framework in place, DOD would be better positioned to reduce its significant reliance on petroleum-based fuel and to address the energy challenges of the 21st century.

In: Defense Energy Management
Editor: Edward R. Myers, pp. 1-91
ISBN: 978-1-60692-574-4

Chapter 1

ANNUAL ENERGY MANAGEMENT REPORT (FISCAL YEAR 2007, JANUARY 2008)

Department of Defense

EXECUTIVE SUMMARY

The Department of Defense (DoD) continues to make significant progress toward achieving the goals of the Energy Policy Act of 2005 (EPAct 2005) and Executive Order (EO) 13423, *Strengthening Federal Environmental, Energy, and Transportation Management.*

The DoD Energy Program initiatives include energy awareness efforts, energy manager training, audit programs, procurement of energy efficient products, and the use of sustainable design in new construction and major renovation. Other contributing factors include integrated energy planning, enhanced use of renewable energy, demonstration of innovative technologies, and the use of Energy Savings Performance Contracts (ESPC) and Utility Energy Service Contracts (UESC).

DoD is responding to EPAct 2005 and EO 13423. Combined, these mandates established a new energy baseline (2003), increased the annual reduction requirement to 3 percent per year, increased the percentage of renewable energy required (7.5 percent by 2013), increased energy efficiency of new construction to 30 percent below the current standard, and required metering electricity consumption of all facilities.

Through Fiscal Year (FY) 2007, the Department of Defense achieved a 10.1 percent decrease in goal facility energy consumption (as measured on a British

Thermal Units (Btu) per gross square foot (GSF) basis [Btu/GSF]) as compared to the revised 2003 baseline. The Department of the Army determined that the square footage was over reported in 2003, compared to data contained in the real property database. Therefore, this chapter contains a significantly revised baseline, which raises the previously reported Btu/GSF from 113,510 to 116,134.

At the end of FY 2007 the Department has 1.95 billion square feet of facilities and spent $3.4 billion on facility energy. DoD spent $9.5 billion on non-fleet vehicles and other equipment – such as auto gasoline, LPG-Propane, Aviation Gasoline, jet fuel and Navy-special fuel.

DoD continues to make progress in installing renewable energy technologies and purchasing electricity generated from renewable sources (solar, wind, geothermal, and biomass) when life cycle cost-effective. The National Defense Authorization Act of 2007 codified a 2005 DoD goal to produce or procure renewable energy equivalent to 25 percent of facility electrical consumption. The total renewable energy that the Department produced or procured in FY 2007 amounted to 12,054 trillion Btu and represents 11.9 percent of the facility electrical consumption. For FY 2007, the Department of Energy revised the guidance for compliance with the renewable energy requirements of EPAct 2005 and EO 13423, allowing only renewable electricity. Under this revised guidance, DoD achieved 5.5 percent total and 3.3 percent new renewable energy, well exceeding the goals of 3 percent and 1.5 percent respectively.

I. Management and Administration

A. Energy Management Infrastructure

1. Senior Agency Official

The Deputy Under Secretary of Defense (Installations and Environment) (DUSD(I&E)) is the DoD Senior Agency Official responsible for implementing the goals of EPAct 2005 and EO 13423.

At the Service and Agency level, the following are senior level officials:

Department of the Army (DoA)

The Senior Agency Official for the DoA is the Assistant Secretary of the Army for Installations and Environment (ASA(I&E)).

Department of Navy (DoN)

The Assistant Secretary of the Navy for Installations and Environment (ASN(I&E)), is the designated senior DoN official for Energy.

Department of the Air Force (DoAF)

The Secretary of the Air Force (SECAF) Memorandum, "Clarification of Roles, Responsibilities, and Authorities Following the Departure of the Undersecretary of the Air Force," September 4, 2007, designated the Assistant Secretary of the Air Force for Installations Environment and Logistics (SAF/IE) as the Senior Energy Official. SAF/IE co-chairs the Headquarters Air Force Energy Senior Focus Group (SFG), providing energy leadership and guidance to the entire United States Air Force.

Defense Contract Management Agency (DCMA) DCMA energy chain of command:

Administrator for HQ DCMA Energy Management Program Energy Manager DCMAA-ACTF, Bratenahl, OH

Energy Manager DCMAW-CS, Carson, CA Energy Manager DCMA-D SEA, Alexandria, VA

Defense Commissary Agency (DeCA) DeCA energy chain of command:

Director

Chief Operating Officer

Director of Performance and Policy

Agency Energy and Environmental Manager

Defense Finance and Accounting Service (DFAS)

The Director of Support Services has been appointed as the agency's senior energy official.

Defense Intelligence Agency (DIA) DIA energy chain of command is:

Director, Defense Intelligence Agency Director, Directorate for Mission Services Chief, Engineering and Logistics

Chief, Facility Engineering Division

National Geospatial-Intelligence Agency (NGA)

The Director, Installation Operations Office is the Senior NGA official responsible for installation operations at all assigned sites.

National Security Agency (NSA)

The Deputy Associate Director for Installations and Logistics, is the designated senior NSA official for energy and oversees the NSA energy program.

Tricare Management Agency (TMA)

The Bureau of Medicine and Surgery (BUMED) Senior Official for energy management is CAPT Stephen Bell. It is BUMED's responsibility to provide staff support and to help its activity stand alone installations and Military Medical Treatment Facilities (MTFs) pursue compliance with all energy mandates. All BUMED Navy Medicine activities follow applicable policy issued from ASN(I&E), ASN Installations and Facilities, and Chief of Naval Operations (N46). BUMED activities also follow guidance issued from Commander, Navy Installations Command (CNIC) and Commandant of the Marine Corps, Facilities and Services Division.

The United States Army Medical Command (MEDCOM) Senior Official for energy management is Mr. Mike Sartori.

Washington Headquarters Service (WHS)

The Pentagon Energy Manager is located within the Defense Facilities Directorate (DFD) in the Engineering and Technical Services Division (ETSD.) The official's role and responsibilities are to provide leadership for the WHS Energy Management Program and obtain resources for energy conservation projects.

2. Agency Energy Team

DUSD(I&E) instituted an EO 13423 working group charged with horizontal and vertical integration of all aspects of implementation. This Integrated Product Team includes members of myriad other working groups to ensure that EO 13423 goals are imbedded across the Department.

The Department Energy Program Manager is also the Chairman of the GovEnergy Executive committee, charged with planning and executing the Federal Government's premier training conference for energy management.

At the Service and Agency level, the following is a description the Agency Energy Teams and some of their responsibilities:

DoA

The Agency Energy Team is comprised of persons from the Office of the DASA(I&H), Office of the Assistant Chief of Staff for Installation Management and the Installation Management Command in collaboration with the US Army Corps of Engineers, Office of the Assistant Secretary of the Army for Acquisition, Logistics and Technology and other Army Offices and Commands.

DoN

The Deputy Assistant Secretary of the Navy for Installations and Facilities, is designated as the Chairman of the Department of Navy Shore Energy Policy Board..

The Director of the Department of Navy Shore Energy Office serves in the, Office of the Deputy Assistant Secretary of the Navy for Installations and Facilities.

The Navy's Shore Installation Energy Action Officer represents CNIC.

The Marine Corps Shore Installation Energy Manager represents Headquarters United States Marine Corps and serves in the Facilities and Services Division Facilities Branch.

DoAF

On September 18, 2007, SECAF issued a memorandum, which outlines the energy strategy of the Air Force and communicates the commitment to meet the goals stated by the President in EO 13423. To accomplish this effort, the Secretary, directed that the energy policies and instructions governing energy management be updated and directed all Major Commands (MAJCOMs) and Installations to create Energy Management Steering Groups (EMSGs) to manage execution of the policy.

The Energy SFG serves as the Senior EMSG within the Secretary and Headquarters (HQ) United States Air Force (USAF). The Energy SFG's scope extends to all energy use and conservation issues within the Air Force, including seeking alternative energy opportunities. This includes energy use at all Air Force installations, within ground transportation and support equipment/systems, aviation fuel use, and associated science and technology opportunities. It designates working groups to focus on specific energy issues within the Air Force. The Energy SFG is co-chaired by the Air Force Senior Energy Official and the Vice Chief of Staff of the Air Force with membership organizations from the Secretariat and Air Staff.

The SFG has four working groups: Provide Infrastructure Working Group, Acquisition & Technical Working Group, Aviation Operations Working Group, Synthetic Fuels Working Group, and three advisory groups: Strategic Communication Integration Working Group, Critical Infrastructure Program Working Group, and the Innovative Financing Working Group. The SFG has approved a 14 point, $240M, facilities energy program for FY 2008 through FY 2013.

On July 14, 2007 the SECAF issued a second "Letter to Airmen" highlighting progress the Air Force has made in energy conservation, and requested all Airmen make energy a consideration in all they do, which is a fundamental element of the Air Force Energy Strategy.

DeCA

Each DeCA region has a Region Energy Task Force Team. These Task Forces meet three to four times per year and continuously look for energy saving programs/projects in an effort to achieve both short- and long-term energy efficiency.

DFAS

The DFAS Corporate Energy Management Team includes the DFAS Facilities Program Manager and the DFAS Facilities Specialist.

Defense Logistics Agency (DLA)

DLA's team includes the Administrator for HQ DLA Energy Resource Management Program and the following positions:

Energy Manager Defense Depot San Joaquin California
Energy Manager Defense Supply Center Richmond Virginia
Energy Manager Defense Supply Center Columbus Ohio
Energy Manager Defense Depot Susquehanna Pennsylvania

NGA

The energy program for NGA falls under the Quality Assurance Division in the Installations Operations Office. The Quality Assurance Division provides support to the sites in the development of operational procedures and assist in the development of site-specific energy and water management programs. The Division also prepares the annual Energy Report. The site managers and their assigned staff develop and implement all site-specific plans and pursue strategies to accomplish energy and water consumption reduction goals. Team is composed of: NGA Energy Manager

Site Manager Bethesda Action Officer Bethesda Site Manager St. Louis Action Officer St. Louis

Site Manager Washington Navy Yard

Action Officer Washington Navy Yard

Site Manager Reston Action Officer Reston

NSA

NSA's team includes: NSA Energy Manager Chief of Facilities Engineering

Lead Mechanical Engineer Lead Electrical Engineer Chief Infrastructure Maintenance

Deputy Chief Operations, Maintenance, and Utilities

Finance and Accounting Finance and Accounting

The NSA Energy Team is responsible for identifying and initiating potential energy conservation projects throughout the NSA campus to meet the objectives of EO-13423.

TMA

The BUMED Energy Team comprises an energy program manager at BUMED and energy managers at the Navy Medicine regions and stand alone installations. At each MTF, the facility manager or a member of his staff, is charged with coordinating energy matters with the host installation's energy manager. Naval Facilities Engineering Command (NAVFAC) provides contracting support for ESPC. TMA provides programming support in developing Energy Conservation Investment Program (ECIP) Candidates. NAVFAC and the Department of Energy Oak Ridge National Laboratory (ORNL) provide additional support for BUMED's Energy Management Program.

Although BUMED has the overall responsibility for Navy Medicine's Energy Program, each of BUMED's activities have individual Energy Programs specific to the activities' stand alone installation or are following their host command's (CNIC or Commandant Marine Corps) Energy Programs. BUMED is in the process of re-establishing its overall Energy Management Program and has awarded a project order to ORNL to perform the following in the next year:

Establish plans for an Energy and Water Use Management Program

Collect energy and water use data and baseline energy and water use performance of key facilities

Determine metering requirements to meet EPAct 2005.

The MEDCOM Energy Team comprises an energy program manager at HQ MEDCOM and energy managers at both installations. At each MTF, the facility manager or a member of his staff, is charged with coordinating energy matters with the host installation energy manager. The MEDCOM Center for Healthcare Contracting provides contracting support for ESPC. US Army Corps of Engineers (Huntsville Engineering Support Center) provides programming support in developing ECIP projects.

MEDCOM established an energy management division headed by the MEDCOM Energy Program Manager. The MEDCOM energy program manager receives policy guidance from DOE, DoD, TMA, and through coordination with HQ DoA and other federal agencies. MEDCOM issues further guidance as necessary and supports the installations and MTFs in executing their energy programs.

The team's mission is to interpret, establish, develop, and implement energy management policy, conserve energy wherever possible, protect the environment, and develop and recommend energy consumption improvements for use throughout MEDCOM's facilities. Energy conservation throughout Army medical facilities and installations is accomplished by improving efficiency and eliminating waste through policy directives, infrastructure improvements, specific energy management actions and installation of meters throughout the Army medical facility inventory. Major Army medical facilities are currently 100 percent metered for electricity, using standard meters. Upgrading to advanced electrical meters and metering of the remaining medical clinics inventory is expected to be completed by 2012.

Fort Detrick and Walter Reed Army Medical Center energy managers control site-specific energy and water conservation programs at their installations. Energy project managers at other Army Installations MTFs control their energy program in accordance with host installation operating agreements and in coordination with US government agencies, DoD, MEDCOM, TMA and other recognized conservation criteria.

WHS

The DFD Energy Advisory Committee members include the Pentagon Energy Manger as the chairman and the Director, ETSD as well as multiple other members including representatives from the Building Operations Command Center, the Pentagon Building Management Office (PBMO), Federal Office #2 (FOB 2), and the Pentagon Heating and Refrigeration Plant (PH&RP). The team's responsibilities are to formulate and execute energy management strategies to meet or exceed the EO 13423 goals and to report progress. Quarterly team meetings are conducted regularly.

The Pentagon Energy Manager is responsible for the utilities supplied to the Pentagon Reservation which includes the Pentagon, the PH&RP, FOB 2/Navy

Annex, the Remote Delivery Facility, the Modular Office Complex and all Pentagon Reservation grounds and parking areas. The Energy Manager is also responsible for the utilities for the Hybla Valley Office Building and the U.S. Court of Appeals for the Armed Forces. These facilities exceed 8 million gross square feet.

B. Management Tools

DoAF

The SECAF directed EMSG at all Major Commands and Installations. These are multi- disciplined groups consisting of individual from environmental, energy, acquisition, transportation, facilities, procurement, legal, budget, operations, and information technology. Each of these groups will keep meeting minutes that track their progress. Installations will submit minutes to the MAJCOMs and consolidated installation minutes and MAJCOM EMSG minutes are submitted to SAF/IE, who then uses the minutes to measure progress against the Air Force Energy Strategy goals and objectives, which correspond to the EPAct 2005 and EO 13423 energy goals.

Model Energy Base Initiative (MEBI). The Air Force designated two Air Force Bases (AFBs), Barksdale and McGuire, to be model energy bases. These bases will showcase energy projects and management practices as these bases implement the Air Force Energy Strategy of 1) Reduce Demand 2) Increase Supply and 3) Culture Change. The model energy bases serve as test beds for new energy technologies and practices that can be disseminated to other bases after the they are proven. Both of these bases are located in strategic areas of the country, Barksdale is located in Louisiana and will emphasize technologies and practices that will reduce cooling demand and McGuire is located in New Jersey, and emphasizes technologies and practices that will reduce heating demand. Additionally, both bases have active flying missions and will showcase technologies and management practices to reduce aviation fuel consumption.

1. Awards (Employee Incentive Programs)

DoA

The Army conducted the 29th Annual Secretary of the Army Energy and Water Management Awards Ceremony at the annual Federal energy exposition and workshop GovEnergy, with the Principal Deputy Assistant Secretary of the Army for Installations and Environment presenting the awards. Among awardees were individuals and small groups representing Army installations of Fort Stewart,

Georgia; Fort Knox, Kentucky; US Army Garrison Hessen, Germany; Fort Detrick, Maryland; US Army Benelux; US Army Garrison Ansbach, Germany; Rock Island Arsenal, Illinois; and US Army Garrison Wiesbaden, Germany. Representatives from Fort Knox and Rock Island Arsenal also received awards at the 2007 Federal Energy and Water Management Awards hosted by the Department of Energy. Local energy award programs were also conducted by Army installations such as at Environmental Quality Control Committee meetings at US Army Garrison Hawaii and through use of the Army Suggestion Program to reward individuals for their valuable energy and water management suggestions.

DoN

The Department annually holds a Secretary of the Navy (SECNAV) energy awards ceremony to recognize outstanding achievement in the efficient use of energy. Eight SECNAV awards were issued to the top installations in the categories of facilities and industrial installations. In addition, installations with an aggressive and successful program were recognized as achieving Platinum (highest) or Gold (second highest) level rankings for their energy programs. DoN installations and individuals also received 8 of 28 awards for Federal Energy and Water Management.

DCMA

DCMA uses the DoE Federal Energy Management Program (FEMP) Annual Federal Energy and Water Management Awards Program to nominate individuals, facilities, or teams that perform exceptional work in implementing EO 13423.

DoAF

- The Air Force participated in the 2007 Federal Energy and Water Management Awards program. The Air Force was awarded eight winners out of 13 submissions. They are:

  - Energy Efficiency/Energy Management, Organization - The Energy Efficiency Team from the 3rd Civil Engineer Squadron, Elmendorf Air Force Base, Alaska.
  - Energy Efficiency/Energy Management, Organization - The Energy Management Team from the 82nd Civil Engineer Squadron, Sheppard Air Force Base, Texas.

- Water Conservation, Organization - The Water Conservation Team with the 92nd Civil Engineer Squadron, Fairchild Air Force Base, Washington.
- Water Conservation, Small Group - The Water Conservation Group from the 3 77th Mission Support Group, Kirtland Air Force Base, New Mexico.
- Water Conservation, Small Group - The Water Conservation Group from the 12th Mission Support Group, Randolph Air Force Base, Texas.
- Energy Security & Reliability, Small Group - The Energy Management Group with the 17th Civil Engineer Squadron, Goodfellow Air Force Base, Texas.
- Renewable Energy, Individual - The Renewable Energy Team Leader with the 90th Civil Engineer Squadron, F.E. Warren Air Force Base, Wyoming.
- Highest Achievement Award, - Gold Award was presented to Kirtland Air Force Base, New Mexico.

- The Air Force submitted nominations for the 2007 Presidential Awards for Leadership in Federal Energy Management program. The Air Force was awarded one winner out of four submissions. The Air Force winner was the "Air Force Energy Senior Focus Group," comprised of Under Secretary of Air Force and Headquarters Air Force leadership.
- The United States Air Force Spring 2007 "You Have the Power" Energy Champion was the Utility Rates Management Team, located at HQ Air Force Civil Engineer Support Agency (AFCESA), Tyndall Air Force Base, Florida, as their nominee and award winner. A poster of the U. S. Air Force Energy Champion was published and distributed by DOE FEMP to major commands and bases for April's "Earth Day" and was posted on the FEMP web site.
- The Air Force Reduced Energy Appreciation Program (REAP) rewards major installations for their ongoing efforts to reduce energy consumption. Purchased renewable energy is not counted as part of this program. The continental United States bases selected this year are Nellis Air Force Base, Nevada and Holloman Air Force Base, New Mexico, both part of the Air Combat Command (ACC). The overseas selection is Elmendorf Air Force Base, Alaska of the Pacific Air Force (PACAF). These bases reduced their energy consumption by 16.1, 12, and 61.2 percent, respectively, from the previous year.
- Several Major Commands and bases have developed programs to provide incentives, awards and funds. Some examples include:

- ACC funded $1million for a base energy award program, recognized bases for meeting goals and/or for improving energy savings over the previous year's performance. Funds are reserved for energy initiatives.
- ACC provided $3.4 million for new ACC Energy Saving Solutions (ACCESS) fund. ACCESS was funded from MAJCOM funds intended to promote innovative 'out of the box' ideas. ACCESS funded 21 projects reducing energy consumption 42 billion Btu, saving $600,000 annually.
- ACC participated in the new Air Force Energy Conservation Award and presented both small group and individual awards.
- ACC also established a major command energy newsletter to disseminate information on events, ideas, and new technology.
- United States Air Forces in Europe (USAFE) provided incentives to base maintenance contractors to reduce energy consumption through an award fee program. Additionally, USAFE bases participate in Host Nation programs such as the Carbon Trust, which is a United Kingdom organization similar to FEMP.
- Little Rock Air Force Base, Arkansas, Energy Manager received Noncommissioned Officer of the Quarter recognition at the group level. Little Rock Air Force Base uses the Air Force Innovative Development through Employee Awareness (IDEA) Program to reward energy ideas implemented on the base.
- Tyndall Air Force Base, Florida, developed a contest/award program with local utility companies providing prizes for best energy conservation suggestion.
- Vance Air Force Base, Oklahoma, established the "Lightening Bolt" individual and group awards for implementing energy conservation practices on base.
- The 611th Air Support Group at Elmendorf Air Force Base, Alaska allocated a higher percentage of the bonus award fee program on Base Operating Support and Operations and Maintenance contracts for energy conservation. The percentage allocation was increased from 3 percent to 10 percent.
- Kunsan Air Base, South Korea, 8th Fighter Wing Civil Engineer Squadron issues quarterly and annual awards for exceptional staff performance, including those achieving costs savings related to energy savings.
- Osan Air Base, South Korea, awards the Mustang Energy challenge coins to outstanding individuals throughout the year. The base drafted and secured approval for the Energy Warrior certificates to reward individuals

for significant contributions to Osan Air Base's Mustang Energy Conservation Program. The very first Energy Warrior certificate was awarded and publicized on 3 Aug 2007 "Airfield control tower team wins energy award."

- Cape Canaveral Air Force Station (CCAFS), Florida:

  - -Mr. Kevin Hooper, the water engineer, won the General Edwin W. Rawlings Award for outstanding achievements in the field of environmental conservation.
  - -The Space Gateway Support (SGS) Energy Eagle Award is presented by the CCAFS Joint Base Operations Support Contractor (BOSC) quarterly to reward associates for their contributions in meeting or exceeding the SGS, National Aeronautics and Space Administration and 45th Space Wing Energy Reduction and conservation goals.

DeCA

- DeCA uses its existing performance awards procedures and on-the-spot awards.
- DeCA normally participates in the Federal Energy and Water Management Awards program.
- The Agency's Energy Management Program Policy Directive incorporates provisions for incentive awards through our existing Improve Defense Commissary Agency's Efficiency and Service (IDEAS) suggestion program and on-the-spot awards.
- DeCA's East, West, and Europe regional offices established their own, separate energy awards program to encourage energy savings and innovation at store and employee level.
- DeCA East presented two energy savings awards for 2007 as follows: Mr. Michael Dunn, Store Director, Camp Lejeune Commissary. Mr. Dunn investigated his commissary's steam bill; the installation agreed it was too high and reduced the bill by $60,000 a year, a 56 percent savings. Mr. Macario Morin, financial technician and member of DeCA East's Energy Task Force, discovered billing errors for three stores resulting in $91,700 in savings.

DFAS

The Agency Energy Management Program Manager was given a Special Act award for his efforts to promote the Agency Energy Management program during this fiscal year.

DIA

DIA uses its existing performance awards procedures and on-the-spot awards.

DLA

DLA uses DOE FEMP Annual Energy and Water Management Awards Program when possible to nominate individuals and facilities that perform exceptional work in implementing Executive Order 13423. DLA will participate in local and regional awards programs when possible.

NGA

NGA rewards its government Energy Team member's through performance awards. NGA incorporates energy conservation as part of the award fee for BOSC at each site.

NSA

Awards are given to employees for outstanding performance, accomplishments, and innovative suggestions related to facilities projects and programs. Awards are also given to individuals or teams where unique initiatives and exceptional performance proved to be deserving of special recognition. These awards frequently involve cost and energy savings ideas that are beneficial to the Agency. NSA did not participate in local or regional award programs in FY 2007.

TMA

As part of ORNL's project order to aid in the establishment of BUMED's Energy Program, BUMED will establish awards programs, performance evaluations, training and education, and showcase facility requirements to ensure the success of the Energy Program.

MEDCOM encourages participation of energy managers in award programs, and solicits installations' nominations for Secretary of the Army Energy and Water Management Awards. Individual MTFs have been recognized with awards through their participation with the host public works activity. In FY 2007, MEDCOM submitted individuals from Fort Detrick, MD as candidates for the Secretary of the Army Energy and Water Management Awards, winning the Small Group award for Water Conservation. Additionally, facility managers at Madigan Army Medical Center earned the DoE Energy Star Award for hospital facilities.

WHS

PH&RP has an Award Fee in the contract as a Most Efficient Organization (MEO) resulting from a previous A-76 Competition. The PH&RP personnel must maintain performance requirements of the equipment through proper staffing, maintenance, repairs, etc. Each quarter the MEO may get up to $25,000 in awards, a portion of which directly relates to energy efficient operations. Through 3 quarters in FY 2007, they received 67 percent of the total available award with the 4th quarter award pending.

2. Performance Evaluations

DoA

During FY 2007 the Vice Chief of Staff for the Army issued a memo directing Headquarters Department of the Army Principal Officials and Commanders to ensure energy consideration are included in the functional responsibilities of their subordinates and that they include energy and water conservation responsibilities in the position descriptions and performance plans of subordinate commanders and civilian supervisors when appropriate. The memo also stresses the need for energy conservation to avoid increasing demands on limited energy resources, prevent the diversion of funds that could be used for other critical Army requirements and to improve energy security, consistent with the *Army Energy Strategy for Installations*.

DoN

The Navy created an agency-wide standard description of the roles and responsibilities of its energy managers. This description defined all the elements of the energy management function at the installation level and the regional level. These roles and responsibilities are used in existing job series to set the performance expectations. Due to constraints on in-house workforce, DoN is increasingly contracting for Resource Efficiency Managers (REMs). REMs are expected to produce energy savings equal to twice their cost, providing at least a 2:1 return on investment.

DoAF

The Department of the Air Force established two new facility energy centers:

- HQ United States Air Force, Office of The Civil Engineer, Asset Management and Operations Division, Energy Management Branch (A7CAE), provides the policy and guidance on facility energy issues and reports directly to the Air Force Civil Engineer.

- Air Force Facility Energy Center located at AFCESA provides the technical guidance and expertise on all facility energy related matters to the Air Staff, MAJCOMs, and installations.
- All base and MAJCOM energy managers have performance statements that include ratings on implementing energy conservation measures to meet federal goals and Executive Orders for their installations and commands. Additionally, all REMs have performance evaluations based on energy performance standards to meet mandated reduction goals written into their contracts. Other examples include:

  - Goodfellow AFB, Texas, energy program is included in the Operations Flight Chief's appraisal
  - Kunsan Air Base, South Korea: In Fiscal Year 2007, the Maintenance Engineer was responsible for Energy Program Management (per Air Force Manpower Standard 44EO), which includes responsibilities for successful implementation of energy conservation and EO 13423.
  - Yokota Air Base, Japan, Energy Manager has been assigned energy conservation goals (including compliance with EO 13423), managing energy projects and contracts to ensure estimated savings are realized from conception through completion, and monitoring all facilities design and construction activities for compliance with energy management measures.

DCMA

Energy Management responsibilities and duties will be included as part of the individual's performance plan.

DeCA

- Energy management provisions are in the performance evaluations of the DeCA Director and Chief Executive Officer by virtue of his requirement to execute the Agency's Strategic Plan and Key Objectives for FY 2007-2013. Key goals of this plan are: preserve and deliver a premier quality-of-life benefit for our customers; transform the workforce to become more agile, knowledgeable, and motivated to provide exceptional customer service; and maintain and communicate the relevance of the commissary benefit through constant innovation and by strengthening our internal governance. DeCA's strategic plan includes goals for improving our facility condition which includes reducing facility and commissary energy use. This is directly related to reducing our unit operating cost. DeCA has accomplished this by improving the infrastructure through reduction of maintenance backlog, executing new construction, modernization, maintenance and repair, and store equipment requirements that incorporate state-of-the art efficiencies.

- Energy management provisions are in performance evaluations of the Chief, Facilities Sustainment Division; Facilities Program Manager; Agency Energy and Environmental Manager; Energy and Environmental Engineer; and field engineers. Region directors have facility maintenance (including refrigeration/heating, ventilating, and air conditioning (HVAC) maintenance) responsibilities in their position descriptions. DeCA West field engineers have references within their respective position descriptions stating that they are to comply with EO 13423 and that will be DeCA West Energy Task Force Team Members. The DeCA field engineers are also identified as the Region Energy Conservation Officer for their respective region. Their responsibilities include review of regional office managed projects to ensure energy conservation and sustainable design practices are being implemented. Energy conservation design practices are also included as duties and responsibilities in the position description of DeCA field engineers. Other critical elements include review of energy use to ensure timely reporting. A critical element in their work performance includes HQ DeCA energy program policies and guidance implementation and reporting.
- Sustainable design practices are included as duties and responsibilities in the position description of the DeCA West field engineers. DeCA West Region Chief of Operations, zone managers, and store directors have a critical element in their performance plan for unit cost management and control. Store level utilities costs are a significant part of store level costs of operations. Utilities cost management, control, and awareness include quarterly reporting of energy use and costs from each store.
- Performance plans of each DeCA region Chief of Operations, zone managers, store directors, and central distribution center (CDC) managers have a critical element for unit cost management and control. Utilities cost are a significant part of facility operational costs; therefore, proper reporting is essential. Utilities management, control and awareness include quarterly reporting of energy use and costs from each facility to the region designated energy conservation representative, who reviews, consolidates, and forwards to HQ DeCA.

DFAS

DFAS has added appropriate statements including successful implementation of conservation actions in the position descriptions and performance evaluations of members of the agency energy team.

DLA

DLA has an Energy Manager at each host facility. The reporting requirements for energy consumption and cost is described as other duties as assigned within the position description.

NGA

- NGA includes energy conservation as part of the government team members' performance appraisal process.
- BOSC team member's performance includes energy conservation as a consideration during contract evaluation for award fee.

NSA

The NSA Energy team is committed to cost effective energy saving projects and programs designed to benefit the Agency in numerous areas. The senior energy official and NSA's Energy Manager have provisions included in their annual employee performance appraisal directed at energy conservation.

TMA

Performance evaluations of the MEDCOM energy management staff use energy-related metrics as the basis of the performance measured by the National Security Personnel System assessments. These metrics measure the effectiveness of the Command's energy manager by establishing goals for implementing energy initiatives, and measuring progress toward meeting those goals. Additionally, the MEDCOM Assistant Chief of Staff for Environment and Facilities Management maintains specific energy objectives/target dates as part of the Command's Strategic Performance Action Plan, and advises the Chief of Staff quarterly on the status.

WHS

Some job descriptions and critical elements include energy conservation principles for appropriate management and operations personnel and are updated annually. In FY 2007, PBMO continued its efforts to review and update job descriptions for appropriate management and operations personnel.

While it did not achieve its goal of updating 10 percent of job descriptions, the PBMO is progressing in implementing this management tool.

3. Training and Education

DoA

The Army saw a large increase in the number of people receiving energy management training in FY 2007. HQ DoA conducted training at the two-day Army Energy Forum attended by nearly 200 people and conducted in conjunction with the annual Federal energy exposition and workshop GovEnergy in New Orleans Louisiana. DoA also sponsored a certified energy manager training course conducted by the Association of Energy Engineers (AEE).

The Army National Guard hosted a two-week training session to train facilities personnel in effective energy management and water conservation, facility design and funding programs. The training covered many subjects such as commissioning, sustainability, water conservation, utility energy cost and consumption tracking, life cycle cost analysis, utility invoice analysis and code compliance. The Installation Management Command (IMCOM) conducted twelve Energy Awareness and Conservation Assessments in FY 2007 to increase energy awareness of installation personnel and assist the installation to identify quick payback energy saving opportunities with low investment costs. Locations at which these Assessments were conducted are Picatinny Arsenal, New Jersey; Fort Hamilton, New York; Fort Myer/McNair, Virginia; Fort Meade, Maryland; Fort Leavenworth, Kansas; Dugway Proving Ground, Utah; Anniston Army Depot, Alabama; Fort Sill, Oklahoma; Camp Zama, Tokyo, Okinawa and Torii Station, Japan; and Camps Walker, Henry and Carroll, Korea. IMCOM also conducted Energy Assessment Training at West Point, New York and an Energy Summit Workshop for high level Army decision makers to increase visibility and focus of Army energy management programs.

The US Army Corps of Engineers (USACE) Construction Engineering Research Laboratory, Engineer Support Center Huntsville and Corps District Offices conducted several energy training courses such as Installation Energy Workshops, ESPC training, and energy efficient design practices to meet EPAct 2005 efficient design requirements. The Corps also conducted training courses through the Proponent Sponsored Engineer Corps Training program such as in sustainable design, electrical design, emergency power generation and boiler operation, maintenance and safety.

Army installations and region offices conducted many energy training events including building energy manager training by IMCOM-Pacific; an ESPC video-teleconference training conducted by IMCOM-Korea; a Water Wise and Energy Smart training program conducted by Fort Huachuca, Arizona; self-help classroom training like that conducted by the Fort Riley, Kansas Public Works Directorate; and other robust energy training programs for energy conservation officers and employees as found at Fort Hood, Texas.

DoN

In FY 2007, 240 personnel received training in areas specified in EPAct 2005. These personnel included: Energy Managers, Energy Conservation Officers, Maintenance Mechanics, Planners, Equipment Mechanics, Facilities Supervisors, Accountants, Admin. Officers, Project Managers, APWO' s, Architects, Environmental Engineers, Electrical Engineers, Division Directors, Controls Mechanics, Civil Engineers, Budget Analysts, Boiler Plant Personnel, Regional Energy Managers, and Utility Engineers. This brings total personnel receiving training to 2,512 since the program began.

The training consisted of specific training opportunities under the specified areas of EPAct 2005: Operations and Maintenance, Controls, Design, Lighting, Electric Codes, Leadership in Energy and Environmental (LEEDTM) Training, Natural Gas Seminars, Water Resource Management, Steam Plant Improvement, Renewable Energy, Energy Accounting, Energy Savings Performance Contracting, Measurement and Verification, Training on Equipment found in Federal facilities and Certified Energy Managers (CEM) Training. DoN has seen 162 personnel registered as "Certified Energy Managers" since the program's inception.

The sources of training include in house and commercially available sources such as: North Carolina University, Dept. of Energy, Association of Energy Engineers, Johnson Controls, Online Courses, Utility Companies, Tempcon, Allen Bradley, Northwest Energy Efficiency Council, Sandia Nat'l Labs, Integrated Electronics Engineering Center, American Institute of Architecture, National Technology Transfer, Inc, Navy Civil Engineer Corps Officer School, American Solar Energy Society, University of Wisconsin, Federal Energy Management Program, The U.S. Green Building Council, American Water Works Assoc., Florida Solar Center, and General Services Administration.

DoN continued and expanded its energy awareness program to train all personnel to be aware of and influence energy consumption. The program includes compact disks that provide policy, publications and program execution tips for energy managers, as well as materials targeted to educate and involve military youth. Quick dissemination of key information is achieved through the distribution of a monthly newsletter titled *Energized,* and weekly flash emails to energy managers, regions, and HQ. Promotional materials are distributed to personnel to involve all in energy management practices without impacting productivity.

DoAF

Training and education is critical to the Air Force Energy Strategy and the energy vision. The Air Force is emphasizing energy training to all uniformed and civilian personnel and has made good progress in training our personnel. This emphasis on training will continue into FY 2008 as energy communication strategies are developed and rolled out. Below are highlights of energy training for FY 2007.

- Twenty-six Air Mobility Command Energy team personnel completed 504 total hours in energy management training requirements at a cost of $29,000. Training included the Air Force Institute of Technology Energy Management Technology and Policy course and the 2007 GovEnergy Conference.
- Air Mobility Command established an Energy Gram that is distributed throughout the command to raise awareness at all levels on energy management and energy reduction.
- The Air Force finalized development of an automated computer power management initiative to put computer workstations into an energy saving "sleep mode" when left on and idle. This program, Monitor Power Management, was deployed to the field in April 2007. Another program, the Computer Box Power Management strategy is ready for field testing. Both programs have been adopted by Air Force Network Operations for use as an interim energy management solution for all Air Force workstations.
- Air Education and Training Command (AETC) installations trained over 1,800 facility managers in energy awareness and energy conservation methods to reduce consumption within their facilities.
- The United States Air Force Academy (USAFA):

o Provided training at monthly base "Newcomers" briefing on USAFA Energy Management Strategy.

o Provided Energy/Water Conservation Briefing at Annual Facility Managers Training. o Assisted in development of Energy Awareness Article with USAFA Research Department.

o Provide USAFA Cadets with posters and give-aways (calendars, note pads, etc.) on energy awareness.

- Hosted Energy Awareness information table at USAFA Annual Picnic.
- Involved in Energy Engineering Research Project with USAFA Instructors and Cadets. Current project involves design and development of photovoltaic (PV) solar renewable energy system for '08 Cadet senior project.

- PACAF initiated monthly energy conference calls where bases share information on lessons learned and what is working and not working. In addition, base energy personnel are frequent contributors of energy articles in base publications and other media. Other base specific examples are:

  - 611th Civil Engineer Squadron, Elmendorf Air Force Base, Alaska - Provided LEED training to all Civil Engineer Squadron officers and engineers and Civil Engineering officers.
  - Andersen Air Base, Guam, trained 160 facility managers to recognize and work to remediate energy wasting procedures and equipment.
  - The Energy Team at Eielson Air Force Base, Alaska, promoted energy conservation awareness through building manager trainings (held twice in FY 2007, 150 personnel trained this year), and in weekly Work Order review meetings.
  - At Elmendorf Air Force Base, Alaska, the base REM attended two LEED training classes, and completed on-line training in advanced metering systems. Also, the base hosted LEED training for Accredited Professionals that included 51 paid participants from around the state and seven Air Force attendees. It was the first such training held in Alaska in several years.
  - 15th Civil Engineer Squadron at Hickam Air Force Base, Hawaii, attended the Hawaii Joint Services Distributed Generation Industry Forum.
  - Hickam Air Force Base, Hawaii, organized a PV training session for DoD and other interested parties to attend. Sessions focused on technical elements such as sizing, designing, and modeling a solar energy system. There were 92 attendees at the PV training.
  - The Kadena Air Base, Japan, Energy Manager took LEED training. In addition, the base worked with American Forces Network radio/television to feature Energy Projects two times, on their radio and television programs.
  - At Kunsan Air Base, South Korea, newcomers received an energy conservation lecture delivered in-person by a trained base energy official.

- o Kunsan Air Base, South Korea, base housing officers received specialized briefings on the requirements for and method of selecting of ENERGY STAR® equipment for dormitory purchases.
- o At Yokota Air Force Base, Japan, the Energy Manager and REM worked with American Forces Network (Radio/television), Stars & Stripes, and the Fuji Flyer on low cost/no cost energy and water conservation measures which were published and aired on television and radio throughout the base community. The broadcasts increased energy awareness and identified low cost/no cost energy and water conservation opportunities to help the base reduce energy consumption.
- o In FY 2007, Yokota Air Base, Japan, provided energy training to 60 percent of facility managers.

- Hurlburt Air Force Base, Florida, accomplished two public awareness campaigns during FY 2007. The campaign utilized articles in the base paper, earth day celebration and local electrical company educated the base on their "Good Sense" program. The 1st Special Operations Wing at Hurlburt Air Force Base, Florida, established a base energy policy letter. The energy policy multiplies energy saving initiatives like thermostats settings, shutting off computer monitors and the elimination of appliances in personal work spaces.

DCMA

DCMA Energy Manager attended the GovEnergy 2007 Workshop. HQ DCMA Energy Management Administrator attended AEE's Course: Theory and Practice of Energy Efficiency & Green Buildings to stay abreast of all new technology in order to implement the provisions of Executive Order 13423.

DeCA

- DeCA energy and environmental engineers and the primary energy consultant attended GovEnergy in August 2007.
- The DeCA Center for Learning hosted eight Facility Energy Supervisor (FES) / Quality Surveillance Representative (QSR) courses during FY 2007. The FES / QSR training course is a 2-day, commissary specific course for the commissary FES / QSR. Instruction includes formal classroom training for energy awareness and conservation practices, quality assurance, reporting, energy monitoring, and the basic operation of refrigeration monitoring and control systems (RMCS). Also, a site visit to a nearby commissary facility is conducted to acquire "hands-on"

experience and serve as a vehicle for student evaluation. The department operations courses address energy usage and conservation in the store director, meat management, produce management, grocery, and quality assurance courses. These courses are taught throughout DeCA's worldwide operations.

- DeCA continues to pursue a goal of two trained FESs per store. The target audience in FY 2007 was commissary store administrators, general managers, and associates responsible for completing DeCA energy management reports. DeCA trained 143 commissary store personnel during our FY 2007 training cycle. Training and travel costs were $247,655. Students represented facilities in Europe, Far East, and CONUS. The FES / QSR students are normally department managers, quality assurance, or store supply personnel.
- All DeCA employees are required to view our 12-minute, commissary energy awareness video, "Put Yourself in the DeCA Energy Efficiency Picture," within 30 days of hire, which is also incorporated in our FES / QSR and executive courses. This commissary specific, energy awareness training video is provided to all DeCA commissaries, CDCs, and office facilities. All associate and store level contractors are also required to view DeCA's Environmental Management System (EMS) Awareness Video distributed in FY 2007.
- Each commissary, CDC, and office facility has an energy management supervisor whose title is "Facility Energy Supervisor." Commissaries and CDCs have a requirement to maintain two FES / QSR trained personnel on staff at all times. This person attends formal training minimally every 2 to 3 years.
- Training materials, including videos, are available at each store and office for initial and refresher energy awareness training.
- Annual continuing education unit (CEU) requirements to maintain state licensing are met by reviewing articles in professional periodicals and attending related training which includes articles in various energy alternatives and conservation. Benefits are maintaining professional licensing and awareness of current and developing new technologies impacting energy conservation.
- In FY 2007, ENERGY STAR products continued to receive utmost consideration when developing specifications and issuing acquisitions for energy using products.

- Information technology hardware and computer and copying equipment are acquired under the ENERGY STAR program using GSA schedules and either government-wide or service contracts.

DFAS

One member of the Energy Management Team attended Distributed Generation Training Program, live web cast, May 2007. Pertinent energy information has been posted to the DFAS ePortal site for employee access.

DIA

- CEU requirements to meet Agency professional development objectives offer opportunities to take classes involving various energy conservation strategies.
- In FY 2007, equipment energy efficiency was mandated in specifications developed when issuing acquisitions for energy using products. All new equipment purchases were made to comply with ENERGY STAR products and American Society of Heating, Refrigeration and Air Conditioning Engineers (ASHRAE) Standard 90.1, Energy Standard for Buildings Except Low-Rise Residential Buildings.

DLA

All Energy Managers were advised to attend DOE and AEE formal, interactive computer and correspondence courses to stay abreast of all new technology in order to implement the provisions of Executive Order 13423.

NGA

- NGA personnel attended GovEnergy in New Orleans, Aug 2007, (2 site supervisory personnel, 2 site operation and maintenance personnel); approximate cost $10,000.
- The NGA Energy Manager attended the following training in FY 07:

  - Leadership in Federal Energy Management, DoE FEMP sponsored training for energy officials; approximate cost $1,000.
  - National Conference on Building Commissioning in Chicago, IL in May 2007; approximate cost $3,000.

NSA

NSA maintenance staff and first line managers attend training for their respective disciplines. Several employees also attended the GovEnergy 2007 workshop. Personnel attend conferences and symposiums, which often include energy conservation modules. Benefits derived from this training include awareness and application of new and emerging energy saving technologies, new techniques, materials, and best management practices.

Also new for FY 2007 is an Energy Conservation Program web page used to notify the workforce about programs and projects. It also has a question and answer section addressing the most commonly asked questions.

TMA

MEDCOM encourages and publicizes various training opportunities for energy managers. This includes energy management training courses available from commercial sources such as AEE and participation in annual conferences such as the World Energy Engineers Conference. Additionally, the MEDCOM Energy Program Manager shares energy information and news through the Command's quarterly news letters. In 2007, MEDCOM held it's first-ever energy conference for all its facility managers and directors. Subjects covered at this conference included facility energy auditing, continuous commissioning, the DOE ENERGY STAR Portfolio, ESPC, and the ECIP. The MEDCOM Energy Program Manager trained and tested for accreditation as a CEM in August, 2007.

WITS

During Earth Week each year, booths are arranged in the Pentagon to educate tenants about energy conservation, recycling and other environmental friendly activities. Pamphlets, cards, and other various reading materials are made available to provide information on practices and their benefits. Also, during October (Energy Awareness Month), booths are set up in the apex to educate tenants about energy conservation.

A FEMP Three Part Training Series on Federal Sector Metering for Compliance with EPAct 2005 was held at the Pentagon. Part 1, Metering Overview included the rational, benefits, technologies, planning, and cost justification of metering. Part 2, Metering Technologies, Communications and Data Storage, included technical aspects of metering equipment, communications options, and data storage. Part 3, Metering Planning, Financing, Uses for Data and Case Studies, included elements of metering plans, options on financing, productive uses for data and Federal sector case studies. Also, Pentagon Renovation Office (PENREN) conducted Sustainability and Environment Integrated Product Team (SE IPT) meetings throughout the year. These meetings focused on initiatives to incorporate best practices in sustainable design and construction practices. These meetings included guest speakers that have knowledge and expertise in the built environmental field to share with PENREN new technologies and industry standards. Twelve of the appropriate personnel types were trained during FY 2007.

4. Showcase Facilities

DoA

The Corps Support Group Barracks Complex at Fort Campbell, Kentucky, has been designated as a showcase facility. Occupancy occurred June 2007. The complex was designed to standards in Engineer Technical Letter 1110-3-491, Sustainable Design for Military Facilities and met sustainable design goals including efficient life cycle use of resources and raw materials, financial stewardship, consideration of renewable resources, work environment and restoration of the natural environment.

Two facilities at Fort Detrick, Maryland, were designated as Federal Energy Saver Showcases in FY 2007. These included projects for lighting controls for the headquarters building, and solar water heating for the indoor pool.

Fort Drum, New York, completed construction of 363 residential community initiative housing units in FY 2007, which were constructed to New York State Energy Research and Development Authority criteria for energy efficient homes.

DeCA

DeCA identified the alteration and lighting study prototype project at Naval Amphibious Base Little Creek, VA, Commissary as the 2007 showcase facility. However, the Little Creek project has been deferred to FY 2009.

Every project DeCA accomplishes (Major and Minor Construction Improvement Program) can be considered a "showcase facility project" because they all encompass the following features to the maximum extent possible. Energy efficiency features include heat reclaim from the refrigeration systems for space and water heating and refrigeration compressor systems comprised of several compressor sizes to ensure the most efficient combination of compressors is running at any one time to meet the load. The RMCS for control of the refrigeration system and HVAC system ensure efficient equipment operation. DeCA maximized use of energy efficient glass door refrigerated display cases, installed automatic scheduling of sales area and display case lighting, anti-sweat heater controls, temperature-terminated defrost, and energy efficient lighting systems. Additional energy efficiency features of this facility are occupancy sensors, automatic water control on restroom fixtures, dual path HVAC systems for the sales area, maximum use of wall and roof insulation, energy efficient doors and windows, and utility metering.

DIA

The designated "showcase facility" for the FY 2008 construction program is the Joint Use Intelligence Analysis Center. A design is currently underway to identify new lighting and HVAC design strategies to reduce the energy consumption to exceed the minimum energy efficiency requirements of ASHRAE Energy Standard 90.1. The proposed design is expected to earn the minimum 33 credits for certification as a LEED Silver building as well as achieve an energy budget of 30 percent less than the baseline building required by ASHRAE Standard 90.1.

DLA

DLA has one showcase facility in the Military Construction (MILCON) Program-construction of a new physical fitness center. This building will include USACE Sustainable Design and Development (SDD) concepts.

II. Energy Efficiency Performance

A. Energy Intensity Reduction Performance

Below is an excerpt from the performance summary of the DoD Energy Data Report

Energy Management Requirement	FY 2003 Btu/GSF	FY 2007 Btu/GSF	Percent Change 2003 - 2007	FY 2007 Goal Target
Reduction in energy intensity in facilities subject to the EPACT and E.O. 13423 goals	116,134.4	104,416.4	-10.1%	-6.0%

FY 2007, DoD achieved a reduction in energy use of 10.1 percent (measured in Btu/GSF) compared to the baseline (FY 2003). The primary energy conservation goal for Federal facilities for the period FY 2006-2015 is a 3 percent reduction per year using FY 2003 as the benchmark. The reduction in energy use can be attributed to increased awareness and energy training, use of renewable energy that is not counted in the total energy use subject to energy reduction goals, recalculation of the FY 2003 baseline from which FY 2007 energy use is compared by importing real property inventory data into the Army energy reporting database rather than using manual input from installations and a general increased effort on the part of the installations to improve efficiency.

1. Goal Subject Buildings

DoA

The Department of the Army made considerable progress during FY 2007 in energy management by implementing the *Army Energy & Water Campaign Plan for Installations.* This overarching, comprehensive roadmap to meet energy and environmental objectives that are mandated by the EPAct 2005 and EO 13423, as well as to support the *Army's Energy Strategy for Installations* and Federal Regulations and guidance, was developed in 2006 and is updated biennially to define Army Energy Program direction and resource requirements. The Army is committed to the reduction of energy waste in existing facilities, increasing energy efficiency in new construction and equipment, reducing dependence on fossil fuels, conserving water resources, and improving energy security. In FY 2007, the equivalent of 7.5 percent of total facility electric use came from renewable energy sources. The Army continued implementing its energy strategy through the use of utility privatization contracts, the ECIP, ESPC, UESC, and conducted an Army Energy Forum and CEM training for Army energy managers.

In FY 2007, while increasing troop strength, the Army was still able to improve energy efficiency to a level of 91.9 million British thermal units per square foot of buildings, a reduction of 8.4 percent from base year FY 2003 consumption levels. The Army reduced consumption of all energy commodities, including electricity, which helped to offset increases in commodity unit costs and other utility obligations such as ESPC and utilities privatization costs.

Army facilities collectively reduced energy use per unit area of gross square feet from 100,261 Btu/GSF in FY 2003 to 91,873 Btu/GSF in FY 2007, a reduction of 8.4 percent. The main reason for the decrease is a general increased effort on the part of installations to improve efficiency.

Most installations and almost all National Guard sites report having implemented energy saving projects in FY 2007, such as lighting improvements, installation of energy management control systems, HVAC upgrades, window and door replacement projects, system and temperature reset, water conservation measures and use of renewable technologies. The improvement also includes a large increase of energy training where several installations are now providing energy training to all energy users on the installation, where many installations have conducted formal energy awareness training to large groups in forums such as the installation theatres and auditoriums, as well as specialized classroom training for facility and utilities managers directly involved in Army energy and water management programs. Distributing the *Army Energy and Water Campaign Plan for Installations* to all installations, issuance of command correspondence such as on Energy Awareness Month, installation site visits and participation in many Federal energy forums have also helped to increase energy awareness and conservation efforts.

Part of the progress can be attributed to the accounting methods by which the Army tracks progress toward energy reduction goals. FY 2007 is the first year that the Office of the Secretary of Defense required TMA to submit an Annual Energy Report directly, rather than including energy data from US Army MEDCOM installations in the Army's Annual Energy Report. This action removed energy data of two energy intense installations, Fort Detrick, Maryland, and Walter Reed Army Medical Center, Washington DC, from the Army's total reported energy, with a resulting apparent efficiency improvement of approximately 1.4 percent. A second issue is the way FY 2003 energy use per unit area is calculated. In FY 2007, the Army started using data from the Army's Real Property Inventory for calculation of energy use per unit area rather than square footage data manually entered into the Army's energy reporting database by installations. This eliminated the duplicative effort of installations reporting building area data into a real property data management system and an energy data management system and created a consistent method for counting facility area across all installations. It also removed data subjectivity of installation energy managers to include or exclude facilities outside of standard accounting procedures as they determined appropriate, such as with some tenants that reimburse for their energy use, however, the effort resulted in a 4.0 percent decrease in FY 2003 building area and comparable increase in FY 2003 energy per unit area baseline.

DoN

DoN reduced energy consumption per gross square foot by 11.5 percent relative to the FY 2003 baseline, exceeding the 3 percent annual reduction required by EO 13423.

Due to energy program initiatives, compared to expenditures in 1985, DoN is avoiding $400 million annually in energy costs, adjusted for inflation.

DoAF

Using the adjusted FY 2003 baseline, the Air Force percent change from FY 2003 is a 17.5 percent decrease.

DCMA

In FY 2007, DCMA's actual utilization represented an 18 percent increase from the baseline. This increase was because DCMA began reporting a second facility that was not included in the baseline. DCMA will revise the baseline to FY 2007.

DIA's energy use declined 5.8 percent during FY 2007.

DLA

In fiscal year 2007, the DLA utilization decreased 1.1 percent decrease from the baseline. The lack of goal attainment is attributable to DLA continuation of modernization initiatives consistent with actions to ensure cost effective business operations. This includes the replacement and/or conversion of high square footage, low energy intensity facilities with low square footage, high energy intensity facilities.

NGA

NGA's energy utilization rate for increased 7.7 percent compared to the FY 2003 baseline.

While energy reduction remains a high priority goal, energy utilization has increased compared to FY 2003 as NGA's mission and personnel strength has grown rapidly. The need to provide vast amounts of graphically intense data on a virtual real time basis requires energy intensive production equipment, storage media and transmission systems. A large portion of this year's increase is attributed to full operation of the Reston 3 facility opened in late FY 2006, which was partially offset by closure of the Dulles facility. In addition, NGA's large data center in St Louis has now reached 50 percent of ultimate capacity with a corresponding increase in energy usage, up from 30 percent capacity in FY 2006. Continued increase in energy consumption is expected for the foreseeable future, at least until NGA is able to fully occupy new, more efficient facilities in FY 2011 and close existing outdated facilities.

NSA

NSA had a net increase of 3.9 percent in energy use intensity from the base year to FY 2007. Because of the nature of its mission, NSA has high-density technical loads in many facilities that continue to grow annually.

TMA

The data reported in this section only represents BUMED's seven stand-alone installations: National Naval Medical Center (NMC) Bethesda, MD, NMC Portsmouth, VA, NMC San Diego, CA, Naval Hospital (NH) Beaufort, SC, NH Bremerton, WA, NH Guam, and Naval Health Clinic Charleston, SC. This data was previously reported with the Department of the Navy, but is being reported with TMA as of this chapter.

BUMED had a net increase of 13.5 percent in energy use intensity from the baseline.

Future progress towards achieving energy reduction goals will depend on implementation of an aggressive Energy Program. The following are specific tasks to be performed by ORNL related to the energy reduction goals:

- Benchmark the energy consumption of BUMED's facilities (via ENERGY STAR) -Identify ENERGY STAR candidates
- Help BUMED obtain ENERGY STAR recognition
- Provide recommendations for energy use tracking and reporting.

MEDCOM installations' reduced energy consumption by 1.5 percent.

Several major factors over the past several fiscal years have impacted energy consumption per square foot at MEDCOM installations. Most notably, MEDCOM has experienced increased energy usage by new temporary modular buildings, and an increased mission tempo. Conversely, the excellent success of MEDCOM's continuous commissioning program has yielded outstanding energy savings at the Walter Reed Army Institute of Research, a major energy user at that installation. That location reflects only a small fraction of the actual energy savings achieved by continuous commissioning. MEDCOM performed the process at numerous MTFs, where the consumption (and savings) is incorporated in the host installations' energy report data.

Future progress towards achieving energy reduction goals will depend on implementation of an aggressive energy awareness campaign, resumption of ESPC initiatives, pursuit of ECIP funding, and the continued success of MEDCOM's continuous commissioning initiative.

In futuret reports, TMA will be reporting on additional facilities as metering is available to adequately determine energy consumption at those facilities.

WITS

Energy consumption increased 3 percent from the baseline.

2. *Excluded Facilities*

DoN

DoN excludes mission critical, concentrated energy use transmitters, simulators, cold iron support to ships, and some private party facilities as authorized by the DOE criteria guidelines. A list of exempt facilities is provided in paragraph IV.

DoAF

The Air Force has identified several exempt facilities. Numerous military family housing facilities are privatized or in the process of privatization with the utility systems belonging to the housing contractor. The housing contractor pays for the costs of the utilities excluding them from the energy goals. Several communication/test lab facilities are also being considered as excluded due to energy intensive loads driven by mission and operational requirements and not influenced by conventional building energy conservation measures. An updated list of excluded facilities is included in Section IV, Data Tables and Inventories.

TMA

- BUMED excluded two facilities in its installation facility inventory, which are detailed in paragraph IV.

3. *Non-Fleet Vehicle and Equipment Fuel Use*

- Auto Gasoline and Auto Diesel consumption figures increased from FY 2006. One potential reason is continued capitalization of ground fuel inventories and subsequent processing of consumption data in Business Systems Modernization-Energy, a computer system used for tracking of consumption data .
- Aviation Gasoline consumption was higher and is most likely due to increased consumption in surveillance aircraft.
- Jet Fuel consumption was higher in FY 2007. In contrast, the cost of jet fuel was lower than FY 2006. This is due to a decrease in the cost per gallon from FY 2006. Increased operations tempo in Iraq may be the reason for higher jet fuel consumption.
- Navy Propulsion (Navy Special Category) is also slightly lower for FY 2007.
- The overall consumption in this area increased 4 percent during FY 2007

DoAF

The Air Force approved using a blend of synthetic and oil-based fuel in its B-52s, is testing it on C-17s and will soon test it for B-1 bombers. Wright-Patterson AFB has played a lead role in the fuel initiative and the testing. The fuel being tested is a 50-50 blend of fuel refined from natural gas and fuel refined from oil. In August, the Air Force certified its fleet of B-52s to use the blend. It is currently in the processing of testing the blend on C-17s, and next week will begin ground testing the blend on a B-1 bomber engine.

4. Reporting Energy Efficiency Funding to the Office of Management and Budget (OMB)

In a related reporting action through agency budget offices, OMB Circular A-1 1 requires reporting on the funding that agencies request to implement energy efficiency measures. DoD prepared its response to this requirement. The three primary sections of information are:

1.1. Identification of Funds for Strengthening Energy Management as Required by E.O. 13423

1.2. Identification of Funds for Strengthening Transportation Management as Required by E.O. 13423

1.3. Identification of Funds for Strengthening Other Management Directives as Required by E.O. 13423

DoD prepared a response to Sections 1.1 and 1.2 but is deferring a response to Section 1.3. Currently, resources required to implement EO 1 3423 are not broken out as referenced in Section 1 .3 of the A- 11, but are built into the Department's overarching budget strategy which includes a wide variety of organizations and functions across the Department. The Department does not track specific EO requirements at this time, but feels resources are available and adequate in the current budget to execute EO 1 3423 requirements. To provide oversight, coordination, and direction for the Department to implement EO 1 3423 and ensure resources are used in a cooperative manner, the Lead Agency Official established a cross-functional, senior DoD EO 13423 Executive Committee from organizations having the functional responsibilities and authorities relevant to all of the EO 13423 requirement and goals. Consisting of Senior Executive Service/Flag level representatives from the Office of the Secretary of Defense energy, transportation, facilities, procurement, acquisition, legal, budget, information technology, and environmental functional areas, and each Military Department and the Components, the Executive Committee is charged with developing the guidance, directives, instructions, and resources necessary to execute the EO.

An example of this is in the Department's Environmental Programs. The FY 2008 Compliance Program includes sufficient resources to enable the Department's day-to-day operations to comply with state and local government enforcement of current environmental laws and regulations. In FY 2008, the environmental compliance program increases by $153.4 million to $1,705.6 million. This increase reflects a price growth of $35.7 million and a programmatic increase of $117.7 million as DoD identifies some additional compliance requirements, including EO 13423 requirements as applicable. The Pollution Prevention Program targets the reduction of hazardous material, solid waste, toxic releases, air emissions, and water pollution at the source. The funding requested for FY 2008 of $129.6 million will support these efforts, as well as implementation of Executive Order 13423. Finally, Technology development is important to meet DoD-unique environmental needs with programs that yield quick results and have high payoffs. The FY 2008 request of $214.6 allows DoD to continue environmental research, development, test, demonstration, and validation activities to provide technologies that result in direct operational savings, mitigate future liabilities, and permit the Department to meet its environmental obligations more cost-effectively. EO 13423 is partly reflected in the FY2008 environmental budget.

These are a few examples of the Department's use of current resources to implement EO 13423. The Department welcomes the opportunity to work with OMB to develop reporting criteria and funding requirements for future EO implementation.

B. Renewable Energy

DoA

The Army did not meet the renewable requirement of 3 percent of total electricity use derived from renewable sources as mandated by the EPAct 2005, only obtaining 2.1 percent of total electric use from renewable sources. The Army met the EO 13423 requirement that at least half of the renewable electricity used comes from sources established since January 1, 1999. When considering renewable energy from all sources including those producing thermal energy, total renewable energy was 7.5 percent as compared to electrical energy consumed in FY 2007.

1. Self-Generated Renewable Energy

DoA

The Army had 30 active renewable energy projects operating in FY 2007. Of the total, 15 were generating electricity qualifying for credit toward the renewable energy goal and nearly all the energy produced was used on-site in Federal Army facilities. The exception was an Army National Guard site at which some of the electricity generated was used on state-owned buildings. The Army retained the renewable attributes of the electricity generated for all but one of these projects, a PV roof project at US Army Garrison, Mannheim Germany. This project was funded by the local municipal electric company, which retained the attributes to meet host nation renewable portfolio requirements. Two new projects were implemented in FY 2007:

Location	Project Description	Capacity
White Sands Missile Range, NM	PV street/parking lot lighting	3.3 kilowatt (kW)
Fort Knox, KY	PV Anderson Pool	2.0 kW

In addition to the projects implemented in FY 2007, the Army obtains a substantial amount of electricity generated from renewable sources from hydropower at Rock Island Arsenal, Illinois; PV panels at Fort Irwin, California; PV and wind power at Fort Huachuca, Arizona; a large PV array at Kwajalein Atoll; and eight separate PV and wind power projects implemented by the Arizona Army National Guard with over 60 kW of total capacity.

The majority of energy obtained from renewable sources by Army installations is thermal energy, which do not qualify toward the renewable requirement of EPA 2005. Sources of this thermal energy are primarily from ground source heat pumps (GSHP) at Fort Knox and Fort Campbell, Kentucky and Fort Jackson, South Carolina; and scrap wood to produce fuel for the boiler plant at Red River Army Depot, Texas.

DoN

DoN is increasing generation of renewable energy, operating the largest wind/diesel hybrid plant in the world and the two largest Federal PV systems in the United States. DON is generating "free" thermal energy from the waste heat of five cogeneration systems, and contracted for a sixth plant in Yokosuka Japan, due to be on line in June 2008. DoN generated 308,431 megawatt hour (MWh) of renewable electricity and 1.8 trillion Btu of renewable thermal energy in FY 2006.

The Navy's geothermal plant at Naval Air Weapons Station China Lake, CA, on average, delivers 1.4 million MWh electricity annually, to the state electric grid. Most of this resource was developed prior to 1990 and does not contribute to the Navy's EPAct 2005 renewable goal.

A 30-year Public-Private venture contract for development of the geothermal resources at Naval Air Station Fallon, NV, was awarded in December 2005. The initial production target is 30 megawatt (MW) with a total allowable development capacity of 160 MW. First production must be on-line by 2013.

Projects made operational in FY 2007 include:

Photovoltaics

State	*Application*
CA	Naval Base Ventura County
CA	Naval Support Detachment Monterey
CA	Naval Air Facility El Centro
JA	Fleet Activities Yokosuka

Solar Thermal

State	Application
CA	Marine Corps Air Ground Combat Center Twentynine Palms

Naval Base Ventura County installed an 87 kW rooftop amorphous silicon thin-film PV laminate system on Building 806 at Port Hueneme. Naval Support Detachment Monterey installed a 10 kW PV system at Herrman Hall which became operational February 2007. Fleet Activities Yokosuka installed PV street lights in November 2006 in the recreational outdoor park area at Ikego housing. Naval Air Facility El Centro installed PV street lights.

Marine Corps Air Ground Combat Center Twentynine Palms installed fourteen solar thermal systems at bathroom and shower facilities.

DeCA

DeCA had no energy usage from self-generated electricity renewable sources in FY 2007.

- Many commissary designs incorporate passive solar features, such as day lighting.
- Heat reclaim is always considered and normally used. Day lighting is also considered in design development.

- GSHP systems are now analyzed when considering design alternatives since larger capacities have become available.
- DeCA initiated installation of a roof mounted, PV array capable of producing an estimated 152 kW at the Los Angeles AFB Commissary, CA. The project was funded through ECIP funds provided to DeCA and the USAF. Construction should be completed in FY 2008.

Air Force

- Dyess AFB, Texas, continues to pursue 5.5 MW renewable energy plant that will generate 44,000 MWh annually from gasified municipal solid waste.
- Cannon AFB, New Mexico, and Davis Monthan AFB, Arizona, installed solar-powered street and security lighting.
- Minot AFB, North Dakota, and Whiteman AFB, Missouri, installed 67 tons of GSHP in the Wing Headquarters facilities saving 5.7 billion Btu annually.
- Moody AFB, Georgia, installed solar water heating for indoor pool reducing natural gas consumption by 50 percent resulting in $20,000 annual savings.
- Minot AFB, North Dakota, awarded 2 GSHP projects totaling $773,000. Projected savings is 9 billion Btu annually.
- Hill AFB, Utah, generated approximately 2.1 MWh of electricity from landfill gas. The project was constructed with planned expansion in mind. A current contract modification is in place to increase plant capacity by adding a 3rd generator, effectively increasing plant production to 3.2 MW.
- Laughlin AFB, Texas, awarded a project for installation of six 1 MW architectural wind turbines.
- Luke AFB, Arizona, installed a PV solar roof on the Base Exchange.
- Eielson AFB, Alaska, refuse derived fuel (RDF) program collected and processed over 560 tons of paper products for use in the base's Central Heat and Power Plant providing 7,820 million Btu of energy and saving 500 tons of coal. The program reduced the waste stream entering the Fairbanks North Star Borough from both the base and the Borough. The program is currently suspended because the pellet plant is inoperable.
- Eielsen AFB, Alaska, has solar panel controller installations at 23 range sites resulting in 13 MWh of production which saved 464 gallons of propane.
- Eielsen AFB, Alaska, has wind turbines at 17 of the remote range sites that resulted in 45.6 MWh of electricity production. The wind turbines saved 1,700 gallons of propane.

- Hickam AFB, Hawaii, employs solar thermal water heating (65 million Btu/year) at C-17 aircraft Maintenance Shop Facility.
- Hickam AFB, Hawaii, awarded a design project for the installation of 400 million Btu/year capacities of solar thermal panels at five buildings on Hickam. The solar thermal panels will offset energy use from propane, diesel, and an electrical water heater.
- Kunsan AB, Japan, has 12 solar powered airfield obstruction lamps installed in locations that are critical to aviation safety, but for which trenching and cabling of power supply lines were impractical. Solar power generated for the operation of these lights amount to approximately 0.2 MWh per year.
- Misawa AB, Japan, has two solar energy systems. One system provides lighting to a remote parking lot and the second provides electricity to rental cabins at the base beach. Total electric generation estimated at 2.0 MWh per year.
- Homestead Air Reserve Base, Florida, produced 1.0 MWh with PV lights installed along a walkway
- March Air Reserve Base, California, produced 8.2 MWh with PV receptors on parking lot lights
- March Air Reserve Base produced 313.1 MWh with a new 0.5 MW PV site.
- Buckley AFB, Colorado:

o Two 10 kW photovoltaic systems were installed on two building rooftops in FY 2007.
o A 1,000 square foot Solar Collector for Ventilation Preheat (Solar-Wall) was installed on one building in FY 2007.

- F.E. Warren AFB, Wyoming:

o The base currently has 1.3 MW (3.8 MWh/year) of wind generation on base.
o A 2 MW wind turbine was funded in FY 2006 with constructions to start in fall/winter 2008.

- CCAFS, Florida, has a PV field. This site:

o Provides continuous power and has a 10 battery backup capability.
o Saves $43,000 by eliminating the need for emergency generator.

- Schriever AFB, Colorado, is currently installing a 3 kW solar array at the Child Development Center.
- Ascension Island in the South Atlantic Ocean provides 3.6 MW of wind and 0.7 MW of solar generated on site.
- Los Angeles AFB, California, awarded an 80 kW solar roof on the Defense Commissary Agency (DeCA) facility with an estimated completion date of May 2008.

DIA

DIA is planning a passive infrared and ultrasonic sensing system for occupancy-based lighting controls in the Defense Intelligence Analysis Center (DIAC) Addition. The open floor plan and generous circuiting of the building make appreciable savings a very realistic possibility. This project is planned for FY 2008.

WHS

The Pentagon has three PV arrays with a combined capacity of 96 kW and smaller PV systems (solar lights) at 48 various locations that combine to bring the total Pentagon Reservation PV capacity to 110.4 kW. This energy source is connected to the PH&RP grid.

The solar thermal system at the PH&RP guard booth consists of 400 square feet of tiles with a total capacity of 11.7 kW to provide lighting, heat and air conditioning for the PH&RP guard booth. It is estimated that 51.2 MWh were produced by the solar thermal system in FY 2007.

2. Purchased renewable energy

DoA

The Army purchased 93,000 MWh of electricity qualifying toward the renewable energy goal, the majority of which in the form of renewable energy certificates (REC) at Fort Lewis, Washington, and Fort Carson, Colorado. Most of the purchased energy that came from renewable sources was from municipal solid waste plants at Redstone Arsenal, Alabama, and Aberdeen Proving Ground, Maryland. The thermal energy does not qualify toward the EPAct 2005 renewable energy goal; however purchase decisions were based on economics, consistent with the *Army Energy and Water Campaign Plan for Installations*, as energy is purchased from these plants at the equivalent average cost of less than six cents per kilowatt-hour. Fort Stewart normally purchases a substantial amount of wood waste as fuel for the central energy plant however the plant was temporarily shut down in FY 2007 during refurbishment. The plant is expected to be back in operation for the winter of FY 2008.

DoN

DoN generated and purchased renewable electricity equivalent to 0.3 percent of annual electricity consumption. These sources include wind energy and solar energy only. Including all renewable energy sources (thermal), DoN is consuming 2.9 percent of its total energy needs from renewable sources.

Air Force

- ACC purchased 347,267 MWh of RECs.
- Dyess AFB, Texas and Minot AFB, North Dakota, use 100 percent renewable electricity.
- Nellis AFB, in public private partnership, leased land to a third party, which began construction on largest photovoltaic array in the Americas on 140 acres of Nellis AFB land. The Air Force will purchase all of the power from the 14.2 MW array. Full operating capability was achieved on November 30, 2007.
- Edwards AFB, California, continued its focus on renewable energy by purchasing over 104,350 MWh of renewable electricity under a separate purchase agreement brokered by AFCESA.
- Robins AFB, Georgia, purchased 4,950 MWh of renewable electricity from Georgia Power.
- Hanscom AFB, Massachusetts, purchased 2,300 MWh of renewable electricity from Hess Electric.
- Hill AFB, Utah, purchased over 480,000 MBtu of steam produced from refuse waste from a local steam supplier. The energy generated was reported as refuse derived fuel in the Defense Utility Energy Reporting System.
- Kirtland AFB, New Mexico, purchased 3,500 MWh of RECs via an AFCESA quantity buy for the Air Force.
- Eglin AFB, Florida, purchased 3,300 MWh of RECs through the AFCESA contracting agreement.
- Tinker AFB, Oklahoma, participated in the Oklahoma Centennial Renewable Energy program and acquired 11,100 MWh of RECs from their electric supplier.
- AETC
- o Purchased 21,000 MW of new RECs
- o Purchased 225,000 MW of old RECs

- HQ, Air National Guard purchased 17,500 MWh of renewable energy through the RECs program, with 8,750 MWh being from new sources
- Minneapolis St. Paul Air Reserve Station purchased 37 MWh of wind power from their electric utility
- HQ, Air Force Reserve Center purchased 3,571 MWh of old RECs.
- Fairchild AFB, Washington, purchased 3,291 MWh of electricity from a renewable source through a local utility company in the Washington Region.
- Grand Forks AFB, North Dakota, purchased 3,324 MWh of electricity from a renewable source through a local utility company in the North Dakota Region.
- HQ, Air Mobility Command, purchased 27,312 MWh of RECs from a new renewable source and 27,312 MWh of RECs from an old renewable source.
- USAFA, Colorado, bought Green Power Blocks provided by Colorado Springs Utilities are from the Ponnequin Wind Facility, Colorado which first went on-line in 1997. In FY 2007, 60 MWh were purchased.
- USAFA, Colorado, purchased 1,500 MWh new and 1,500 MWh old RECs from Sterling Planet, Inc.
- Schriever AFB, Colorado purchases 3.15 MW of RECs each year.

TMA

In FY 2007, Fort Detrick produced and consumed 97,000 million Btu of renewable thermal energy in the form of a waste to steam generation plant.

C. Water Conservation

EO 13423 requires all Federal agency heads to develop a water-use baseline based on FY 2007 water consumption. In FY 2007, The DoD established procedures to accurately report potable water usage, its cost, and the facility area associated with the water usage. DoD estimates that it consumed over 116 billion gallons of water at a cost of over $358 million in FY 2007. The area associated with that usage was over 1,952 million square feet. Following are examples of the actions that DoD services and agencies are doing in implementing water conservation practices.

DoA

In FY 2007, the Army used 45.2 billion gallons of potable water at a cost of $63.5 million. The Army's total water use and disposal have declined for many years. Greater treatment and testing requirements imposed on water suppliers by the Safe Water Drinking Act of 1974 (and amendments) have increased the cost of providing potable drinking water. Similarly, the vulnerability assessments of installation water supplies, emergency response plans, and protective measures required by the Public Health Security and Bioterrorism Preparedness and Response Act of 2002 (PL 107-188) have added to the cost.

Water conservation efforts are environmentally responsible and can help installations stretch dwindling operations and maintenance dollars. Any water conservation measures that reduce wastewater quantities also provide additional cost avoidance.

Many Army installations have reported lower water consumption in FY 2007 due to the privatization of on-base housing to Private Partnerships or missions changes due to the effects of the Base Realignment and Closure Act of 2005.

Army installations in areas of the United States affected by drought and those with water restrictions, naturally have reported much lower water consumption figures for FY 2007. Rain water collection and reuse, a matter of course for the Army National Guard in the Virgin Islands, is being looked at on the U.S. mainland, where LEED credits can be earned for new construction projects.

Many installations have installed water conserving toilets and urinals, low flow faucets and showerheads. The have instituted aggressive leak detection surveys and followed up with repair programs of leaky valves and damaged pipelines which has significantly reduced water consumption by up to 20 percent at one location.

Although no longer mandated by Executive Order, a number of installations continue to follow the Best Water Management Practices, espoused by DOE.

At Army's industrial plants, despite higher production rates due to the Global War on Terror increasing the need for water for processes and cooling, plant managers have installed equipment coolers that reduce the use of potable water, tightened up the condensate return systems on steam boilers to reduce the need for fresh water. At Anniston Army Depot, they have converted the existing paint booths from a water-wash filtering system, to a dry filter system which uses no water.

DoN

DoN consumed 34.6 billion gallons of water at a cost of $80.8 million, establishing a water intensity baseline for future reports at 73.5 gallons per square foot.

DoAF

The Air Force consumed 34.2 billion gallons of water in FY 2007. This is a reduction of 5.1 % from last year's consumption. The Air Force has been tracking water reduction since 2000 with an overall reduction from that time of 34 percent or 17,693 million gallons saved. The Air Force will continue its aggressive water reduction program to meet the new 2 percent per year mandate. Here are some examples:

- Nellis AFB, Nevada, awarded a xeriscaping project, reducing water consumption 45 million gallons annually – 15 percent of entire base usage – and significantly reducing grounds maintenance while improving aesthetics.
- Beale AFB, California, awarded a project to use reclaimed wastewater for irrigation. Using reclaimed wastewater results in a savings of 70 million gallons annually.
- Offutt AFB, Nebraska, installed a well pump to use non-potable ground water for irrigation of the parade field.
- The HQ Air Force Reserve Command has Water Management plans for all nine bases. Eight bases have implemented at least four Best Management Practices.
- Lackland AFB, Texas, uses air-cooled chiller systems to reduce cooling tower usage, chemical water treatment program to reduce blow-downs on cooling towers, and includes water conservation and energy conservation measures in all ESPC projects. The base is increasing water conservation awareness by publishing articles in the base newspaper. In addition, purchased 100 million gallons of recycled water.
- Tyndall AFB, Florida, use of ESPC and water awareness programs have resulted in an FY 2007 annual reduction of 75 million gallons of potable water.
- Kadena Air Base, Japan, awarded a $1.2 million ECIP project to install water saving devices in 17 dormitories. Kadena also awarded a $231,000 Productivity Investment Fund project to replace washing machines in 17 dormitories and 12 lodging facilities, with energy and water efficient models.
- Misawa Air Base, Japan, in FY 2007:

  - In-house crew replaced two failed expansion joints on the central boiler plant steam loop, condensate line. This resulted in a water savings of 13 million gallons per year and water and sewage cost savings of $54,000.

- The monthly review of water meter readings revealed that water consumption for one building was well above historical levels. The remedy resulted in annual water savings are 1.92 million gallons and $8,000.

- HQ Air Mobility Command, purchased over 100 water meters with Headquarters Air Force end of year funds to monitor water consumption and to assist further with necessary adjustments to water management plans.
- CCAFS, Florida, in FY 2007 funded four water conservation projects. The total cost of all projects was more than $6 million. The projects included: o Repair Water Lines, Industrial Area, $2.5 million o Replace Water Service Lines on Armory Road, $293,000 o EC Corrective Action, CCAFS Water Systems, $740,000 o Repair Pump Station #7, $2.6 million

DCMA

DCMA water usage in FY 2007 was 3.3 million gallons at a cost of $10,500. Compared to the previous year, which was 3.4 million gallons at a cost of $10,100. This represents a 1 percent reduction in actual water utilization.

DeCA

- DeCA potable water use and cost for FY 2007 is 319.7 million gallons and $977,700, respectively. The water intensity value is 18.3 gallons per square foot, which is a 3.8 percent decrease from FY 2006 consumption.
- DeCA has completed Water Management Plans at 46 percent or of reporting locations.
- The DeCA design criterion requires low consumption toilets and urinals with electronic flush sensors for new and renovated commissaries. Electronic sensor control valves are specified on hand wash lavatories. At locations where host installations maintain "waterless" urinals, the projects may include the "waterless" urinals.
- DeCA construction projects implemented are required to include low-flow devices for flush valves and lavatory faucets. The use of electronic valves was included in projects to reduce water use.

- All regions continue to upgrade restroom facilities by fitting them with low-flow type fixtures and where applicable, sensor-activated faucets and flush valves. This is an on-going process.
- More emphasis has been placed on the immediate necessity to repair leaky and/or faulty plumbing fixtures as they are identified. If the installation/base maintenance workforce is not available to resolve the problem, outside (off-base) maintenance will be sought to acquire immediate correction of the problem.
- Closer monitoring and improved quality checking of commissary store quarterly water use and cost reports (DeCA Form 20-1) has been implemented. Store directors continue to stress the importance of conserving water in their daily operations.
- All new restroom upgrades within the region facilities have incorporated or are planned to change-out fixtures and fittings with low-flow type, sensor actuated operation units.
- Proposed landscaping for new facilities is closely reviewed during all phases of the design for low-maintenance and watering requirements.
- The DeCA East Utilities Task Force directed that all locations conduct a comparative reading of water meters each month to validate consumption as reported by host installation.
- Landscape irrigation, at times, has been substantially reduced at the DeCA HQ facility, Fort Lee, VA
- Proposed landscaping for new facilities is closely reviewed during all phases of the design for low-maintenance and watering requirements.

DFAS

Water consumption was 5.0 million gallons for FY 2007 with a cost of $26,600.

DIA

- DIA potable water usage for FY 2007 was estimated at 42.9 million gallons.
- The DIA design criterion requires low consumption toilets and urinals for projects involving new plumbing fixtures.
- Landscape irrigation is kept to a minimum. The automatic sprinkler irrigation system was removed during the construction prior to FY 2003.

GA

- In FY 2007 NGA consumed 98.7 million gallons of water at a cost of $623,000, a reduction in consumption of 3.8 % compared to FY 2006. The consumption decrease is attributed to improvements in metering and implementation of Water Management Plans at NGA sites. The resulting consumption intensity of 35.9 gallons per square foot is established as NGA's baseline for EO 13423 reporting purposes.
- NGA established a Water Management Program in FY 2003 which incorporated Water Management Plans for each of our 6 major sites. For FY 2007, NGA continued to place emphasis on implementation of Best Management Practices (BMPs) as an efficient and cost effective way to achieve water conservation.

NSA

In FY 2007 NSA paid $1.5 M for 416 million gallons of water. This is a 2 percent reduction in water use from FY 2006. NSA continues to implement water-conservation measures, and uses water saving fixtures in lavatories during renovation or repair projects. There are several buildings on Fort Meade Post that are included as part of the NSA's total water usage. These buildings are tied into the NSA water distribution system but do not have separate metering to determine water use. Future plans include installing separate water meters on each building to determine actual water consumption.

TMA

In FY 2007, BUMED established a usage of 61.9 gallons/gross square foot rate. As part of ORNL's project order, the following will be established to support BUMED's Water conservation efforts in FY 2008: Benchmark the water consumption of hospitals and provide recommendations for water use tracking and reporting.

In FY2007, MEDCOM water consumption totaled approximately 890 million gallons. Fort Detrick continued its successful leak detection program in 2007, and their consumption for that year was 13.5 percent below the FY 2003 level. Additionally, in FY 2007, Fort Detrick awarded a $3.3 million contract for replacement for new incinerator scrubber system, which will result in calculated water savings of 9.5 million gallons per year.

In FY 2007, both Fort Detrick and Walter Reed Army Medical Center completed Water Management Plans in compliance with EO 13423, and according to the associated DOE guidance. In addition, Fort Detrick has implemented at least four BMPs for all facilities on that installation.

WITS

In FY 2007, the Pentagon consumed 149 million gallons of water resulting in a cost of $790,700 compared to FY 2003 with a consumption of 131 million gallons of water in a cost of $278,300.

D. Metering of Electricity Use

DoA

The Army achieved 100 percent of its agency metering plan milestones in FY 2007. Those milestones were to 1) prioritize Army installations for central funding of installing meters in compliance with EPAct 2005 requirements, 2) develop a performance specification for metering equipment as the standard to be used on all centrally funded installations, and 3) develop an acquisition plan for procuring equipment and services to execute the Army Metering Implementation Plan. As of FY 2007, Army installations report a total of 12,667 standard electric meters in use for individual buildings. This represents approximately 12 percent of the buildings in the Army inventory in which 17 percent of the Army's electricity is used. Projections for installing new meters in FY 2008 are for advanced meters to be used exclusively. These will supplement existing meters rather than replace standard meters. Based on estimates in the Army Metering Implementation Plan, the number of advanced electric meters is projected to increase to 3,842 in FY 2008.

DoN

DoN has installed electrical meters on over 13,000 buildings. 760 of these meters meet the requirements of "advanced metering" as required by EPAct 2005. These facilities account for 53 percent of all the electricity consumption within the Department (4 percent of which are covered by the advanced meters).

DoN has established an Advanced Metering Infrastructure (AMI) Program and is currently performing a global evaluation of all sites and developing a full site design for a prototype deployment. The initial AMI architecture will be designed to deliver energy and utilities benefits from the building, through the meter, via a communications backbone to a data collection center. The first deployment will be built at Naval Base Ventura County and is scheduled to be operational by the spring of 2009. Subsequent deployments will be delivered that build upon the initial design and leverage and lessons learned at Naval Base Ventura County. Deployment plans will be sequenced with a goal of completing by 2012 to comply with EPAct 2005 requirements. The DoN AMI network will be built to accommodate future functionalities and as such will certainly use solid-state internet protocol-enabled advanced meters that read at 15-minute intervals and report at least daily.

DoAF

The Air Force developed a metering strategy for meeting the EPAct 2005 mandate of installing electric meters on all facilities by October1, 2012, that are economically life-cycle cost effective. For FY 2007, The Air Force Civil Engineer provided end-of-year funds in excess of $17 million. Based on the meter data call of April 2007, this expenditure for automated meter readers will provide for 95 percent of the required facilities to be metered and over 80 percent of the electric load. Some examples of implementation include:

- HQ Air Mobility Command received and executed $4.6 million in end-of- year funds for advanced metering.
- HQ Air Force Material Command, received over $7 million in end-of-year funds for advanced metering upgrades at four additional installations and additional work at Robins AFB, Georgia, and Tinker AFB, Oklahoma.
- USAFE used $1.1 million in year-end funds to purchase advanced metering at six installations.
- AETC:

  - Lackland AFB, Texas, awarded a contract to install 118 electric smart meters
  - Randolph AFB, Texas, installed seven electric and three gas automated meters and purchased automated meter reading equipment for 60 percent of facilities.

- Eielson AFB, Alaska, received $260,000 in end-of-year funds to upgrade the existing, nonfunctional automated meter reader systems. The funding will also be used to install new meters base-wide.
- Hickam AFB, Hawaii, automated meter reader project is planned in three phases of construction: Phase I (FY 2007) award for approximately 100 meters and communication system. Communication system includes antennas, server, and a laptop computer.
- USAFA awarded a metering project for 4 million square-feet of facilities at a cost of $580,000.
- The Air Force Space Command received $4 million in Air Force and Air Force Smart Operations for the 21st Century funding to procure meters at FY 2007 year end.

DeCA

87.9 percent of DeCA facilities have electric meters.

The DeCA metering plan is being revised to coordinate additional metering and advanced metering with individual host installations as they pursue their DoD metering plans. The goal is to reimburse the host installations for meters installed with the understanding that DeCA will have real time, electronic access to metered data for energy monitoring and analysis purposes.

DFAS

All DFAS buildings are metered.

NGA

All NGA sites have standard meters providing consumption information for electricity, natural gas and water. Electricity consumption for the NGA College at Fort Belvoir is reported through the Army. Beginning in FY 2008, NGA will audit the two western facilities (Arnold and St Louis) with the intent of replacing existing meters with ones meeting FEMP standards for advanced metering. With completion of the New Campus East facilities in FY 2011, all NGA locations will be served by advanced meters.

NSA

NSA is in the process of updating, replacing, and repairing electric meters around the campus. This work was not complete in FY 2007, but is ongoing into FY 2008 and beyond. Meter installation is scheduled for completion around FY 2014.

TMA

136 facilities at the seven stand-alone BUMED facilities are metered. BUMED met its milestone for 2007 to award the project order to ORNL. The following will be established by ORNL for BUMED with respect to metering in FY 2008:

- Determine the cost-effectiveness cutoffs that mandate metering for BUMED facilities. Using these, ORNL will identify all BUMED buildings that must be metered, and provide recommendations on how these facilities should be metered.
- Assess existing BUMED metering and the level of difficulty for installing new electric meters where they are required. ORNL will also provide BUMED a set of electric meter specifications required to ensure metering is in compliance with EPAct 2005.

Currently, approximately 80 percent of the MEDCOM's facilities are metered for electricity, using standard metering methods. As a major electrical consuming tenant on Army installations, MEDCOM is a stakeholder in the Army Metering Plan, and we intend to install advanced electrical meters through that funding avenue. When cost effective and practical, advanced electrical meters will also be installed through Defense Health Program funded renovation and repair work.

WHS

In FY 2007, all three of the buildings in this chapter are metered. The PH&RP and Wedge 2 of the Pentagon have advanced metering. This advanced metering is a part of the Pentagon Renovation. The amount of advanced metering within the Pentagon will increase as the renovation progresses through each wedge. In the near future, we will be harvesting the data from these advanced meters for trending analysis in order to see any changes in energy usage that may indicate such issues as needed repair and maintenance.

E. Federal Building Energy Efficiency Standards

DoA

USACE is working closely with DOE and the Office of the Assistant Chief of Staff for Installation Management to develop design guides for implementing building efficiency standards mandated by the EPAct 2005. USACE has developed prescriptive design guides for battalion headquarters buildings, permanent party barracks, training barracks, and tactical equipment maintenance facilities, four of the most prevalent types of buildings being constructed in conjunction with Army troop stationing actions. Additional design guides are being developed for dining facilities, company administrative facilities, Army reserve centers and child development centers. Use of these design guides will result in meeting the requirement for new building designs to be 30 percent more energy efficient than ASHRAE Standard 90.1-2004.

The requirement to meet EPAct 2005 design requirements has been in the model Request for Proposal (RFP) for contracted building design and construction services processed through USACE since May 2006. When using the model RFP for a project the designer of record is required to comply with designing buildings to use 30 percent less energy than ASHRAE 90.1-2004, purchase ENERGY STAR equipment, install metering and specify premium efficiency electric motors. This includes barracks buildings which typically would be designed to the International Energy Conservation Code standards for low-rise residential buildings. Although buildings for the Active and Reserve Components are designed to use 30 percent less energy than ASHRAE 90.1-2004 standards, the current standard for National Guard buildings is to meet the U.S. Green Building Council (USGBC) LEED silver standard, due to the mix of federal and state responsibilities and ownership of National Guard facilities.

A list of Army MILCON projects in the FY 2008 program showing compliance with EPAct 2005 requirements for designing new construction projects to be energy efficient is in the accompanying Data Report.

DoN

DoN is in the design process on several new facilities. These projects, due to the timelines involved with the MILCON program, were all initiated (and budgeted for) prior to the established requirements of EPAct 2005 and EO13423. DoN policy is now in place that requires compliance with the EPAct 2005 and EO 13423 for all new buildings beginning in FY 2009.

A listing of the new facilities to be constructed in FY 2008 – none of which meet the 30 percent less than ASHRAE standards, for the reason outlined above, is in the chapter.

DoAF

The Air Force FY 2007 facility designs for new facilities (MILCON) reports 51 out of 77 submitted are at least 30 percent more efficient than required by relevant codes, such as the ASHRAE Standard 90.1 or the International Energy Conservation Code, and life-cycle cost-effective. A complete listing of all new Air Force buildings owned, operated, or controlled by the Air Force, for which designs were started in FY 2007, is included in the chapter.

DeCA

- Energy efficiency has historically been given careful attention during the design and construction of commissary facilities. The great energy consumption requirements of commissary equipment has inspired DeCA to pioneer such technologies as heat reclaim and desiccant dehumidification years before the private sector began using them. The DeCA Design Criteria has been updated to incorporate the latest guidance required by EPAct 2005 and EO 13423.
- The DeCA Design Criteria Handbook (DeCAH 20-1) emphasizes life-cycle cost analysis of systems for different types of fuel sources. Designs include other items such as: occupancy sensors; energy efficient lamps and ballasts; LED exit signs; high efficiency motors on air handling units and display cases; maximized use of glass door refrigerated cases

instead of open cases; use of RMCS for the most efficient operation of Refrigeration Systems and HVAC; automatic water controls for restroom fixtures for efficient use of water; use of dual path and desiccant air handling units for the most economical means of cooling and dehumidification; maximized use of wall and roof insulation; implementation of energy efficient doors and windows; and plastic curtains on refrigerated cooler doors.

- Design began on two new commissaries in 2007, Fort Bliss, TX, and Saratoga Springs, NY. They are both expected to exceed the ASHRAE 90.1 requirements by at least 30 percent.

DLA

All of DLA design and construction of new facilities are done by the USACE or NAVFAC, which uses ASHRAE Standard or the International Energy Conservation Code, as appropriate, if life cycle cost-effective.

NGA

In FY 2007 NGA began design of a Base Realignment And Closure project consolidating all sites in the east into a new campus located on the Engineering Proving Ground on Fort Belvoir, Virginia. The campus is being designed to meet the energy reduction guidelines of ASHRAE 90-1.

TMA

Design work for BUMED's new construction and renovation work is accomplished through NAVFAC, who will ensure all new design work will meet those standards where achievable.

Design work for MEDCOM's new construction and renovation work is accomplished through Army Corps of Engineers contracts. MEDCOM has briefed those design agencies of federal energy efficiency and sustainability standards, and all new design work will meet those standards where achievable. Exceptions to these standards will be the use of modular facilities, which the Command has found necessary to meet temporary or sudden mission requirements.

III. Implementation Highlights of FY 2007

DoA

The Army will continue to use the *Army Energy and Water Campaign Plan for Installations* as the means for carrying out the *Army Energy Strategy for Installations*. The *Plan* will be reviewed and updated every two years to set the direction of the Army Energy Program and help determine resource requirements to meet energy goals based on the most recent Federal and DoD mandates and current global energy situation.

DoAF

The Air Force will implement and communicate its strategy and vision during FY 2008 through issuance of an Air Force Energy Program Procedural Memorandum and updating its energy policy, Air Force Policy Directive 23-3 *Energy Management.*

The Air Force has developed an energy strategic plan with three major focus areas 1). Reducing Demand, 2) Increasing Supply, and 3) Culture Change. The Air Force Energy Strategic Plan is a component of and supports Air Force priorities, which are: 1. Winning the War on Terror 2. Developing and Caring for Airmen and their Families, and 3. Recapitalizing and Modernizing our Aircraft, Satellites, and Equipment. The Energy Strategic Plan is the management mechanism that will enable the Air Force to meet or exceed all Federal goals established through public law or Executive Order, including, but not limited to the EPAct 2005 and EO 13423, as well as other mandates of the President and the Office of the Secretary of Defense.

The overarching vision of the Air Force Energy Initiative is "Make Energy a Consideration in All We Do." The Air Force uses energy awareness to keep all personnel focused on energy conservation and efficiency to reduce energy costs. It is only by involving everyone in the Air Force at all levels that the Air Force energy goals will be met or exceeded.

A. Life-Cycle Cost Analysis

DoA

Life-Cycle Cost Analysis (LCCA) was nearly fully implemented for all state Army National Guard energy projects. Army garrisons also made LCCA an integral part of the design process, consistent with the *Army Energy and Water Campaign Plan for Installations*, and options for the best equipment or systems are always evaluated using this tool. Unfortunately, funding is not always available for generally high first-costs that tend to accompany the most cost-effective energy-efficient alternatives when considered on a LCCA basis.

DoN

All DoN energy projects (centrally funded and financed) are required to evaluate savings on a life cycle basis. Projects submitted utilize the National Institute of Standards and Technology publication handbook 135 and DOE energy discount factors as guidance. In FY 2003, the DoN energy projects team adopted use of DOE's Building Life Cycle Costing software as a standard for determining project economics. Sustainable development projects use life cycle costing methodology and follow the whole building design guide. FEMP, ENERGY STAR, GSA and DLA guidance on purchasing energy efficient products continues to be distributed in order to educate purchasers of the life cycle costing requirement and provide them assistance making purchasing decisions.

DoAF

LCCA was used on all new Air Force construction projects and retrofits projects, including ESPC, UESC, and ECIPs. The Air Force is committed to performing annual Facility energy audits on at least 10 percent of the Air Force inventory. Examples include:

- ACC invested $3.4 million in their ACCESS program that requires projects be funded based on best Savings-to-Investment Ratio (SIR).
- Randolph AFB, Texas, Energy Management staff and Facility Managers completed 15 percent of facility energy audits in FY 2007.
- The Air Force Civil Engineer is the infrastructure and vehicle lead on the MEBI to include McGuire AFB, New Jersey, and Barksdale AFB, Louisiana. This initiative will evaluate and implement various sustainable and energy related strategies that will be the template for all other AFBs to execute to meet the sustainable and energy reduction mandates.
- At Patrick AFB, Florida, a LCCA was performed to determine the feasibility of installing GSHPs in military family housing and temporary lodging facilities. The analysis yielded a $2.4 million project with a SIR of 1.82 and energy savings totaling $154,000.

DCMA

A recommendation from the SAVEnergy Audit completed in FY 2006 was evaluated to determine the feasibility or potential for energy/water conservation and renewable energy measures that are life cycle cost-effective.

DeCA

DeCA Design Criteria Handbook (DeCAH) 20-1, emphasizes use of life-cycle cost requirements in the design of commissaries, was revised in FY 2006 and is available on line at http://www.decafacilities.com/decadesign/. The handbook emphasizes life-cycle cost evaluation of HVAC systems for alternate fuel sources and other energy reduction strategies including direct expansion and gas-fired systems. Designs include occupancy sensors, energy efficient lamps and ballasts, LED exit signs, high efficiency motors on air handling units and display cases, maximized use of glass door refrigerated cases instead of open cases, use of refrigeration monitoring and control systems for the most efficient operation of refrigeration systems and HVAC, automatic water controls for restroom fixtures for efficient use of water, use of dual path and desiccant air handling units for the most economical means of cooling and dehumidification, maximized use of wall and roof insulation, implementation of energy efficient doors and windows, and plastic curtains on refrigerated cooler doors. Plastic or metal swing air curtain doors are replacing the strip curtains on walk-in refrigeration equipment in the freezers, meat, dairy, and produce rooms in new and renovation projects. The boiler replacement project at Grafenwoehr Commissary, Germany, used the life-cycle analysis method to determine the benefit for various system components. The use of life-cycle cost alternatives is a primary focus in the design determination required during major and sustainment project development. In FY 2007, an analysis was conducted relative to fax machine repairs at the Agency's Headquarters. Machines with the lowest life-cycle costs were selected for procurement. Submittals during the design phases, including life-cycle analysis of all major construction projects are reviewed and comments are provided by the region as applicable for the specific location.

DIA design criteria emphasizes use of life-cycle cost requirements in new facility construction. Life- cycle costs are evaluated in selection of HVAC system design strategy, such as in selection of distributed water-source heat pumps versus central station air handling systems. DIA design guidance places emphasis on compliance with the EPAct 2005 in achieving 30 percent greater energy efficiency than the baseline ASHRAE-recommended energy requirements (Std 90.1-2004). During major design efforts, consideration is given not only to energy-efficient building envelope design, but also to high efficiency chillers, premium efficiency fan and pump motors, heat recovery chillers, condensing boiler and low-temperature heating hot water systems, water-side economizers, variable water flow chiller evaporators, lighting controls and energy efficient lamps and ballasts.

DIA conducted an energy audit through a utility in FY 2007. This audit will result in potential annual energy savings and energy-related maintenance guidance after recommendations are executed.

DLA

LCCA is used by USACE and NAVFAC for design of new construction and renovations for all DLA projects

NSA

Energy audits identify life-cycle issues. NSA maintains this list of potential energy projects, which are prioritized by the Energy Team and Facility Planning Board according to mission, pay back, and operations and maintenance cost value. New projects are added to the list and prioritized accordingly.

TMA

- Develop a method for targeting BUMED facilities for energy audits, develop a method for targeting facilities for water audits, and make recommendations for an energy auditing process.
- Conduct a demonstration retro-commissioning project at one BUMED facility. The purpose of retro-commissioning is to fix operational problems upfront before other, capital-intensive energy conservation measures are evaluated. This should be done consistently but in fact, is rarely done in practice. It is often focused on energy management systems and direct digital controls. The effort will demonstrate retro-commissioning potential and confirm its need to become a standard practice in BUMED's energy program.

Since 1998, MEDCOM has relied exclusively on ESPC and UESC to perform facility energy audits, and both Walter Reed and Fort Detrick have surveyed most of their facilities through these avenues. Performance of energy audits ceased when ESPC authority expired, and MEDCOM has initiated no new ESPC task orders since that authority was re-instated. As an interim substitute, MEDCOM has aggressively pursued its continuous commissioning program; although not a full facility audit per se, this program does provide a thorough assessment of building mechanical and control systems, and helps to identify system improvements and retrofits. MEDCOM recently developed a prioritized list of candidate facilities in which to perform energy auditing, and awarded a contract task order to the Texas Engineering Experimental Station (a subsidiary of Texas A&M University), to perform an energy audit at Darnall Army Community Hospital, Fort Hood, TX. The site visit for that audit is currently (November 2007) underway.

MEDCOM installations prefer to base their procurement decisions on LCCA, as this strategy offers other advantages in addition to energy savings. In FY 2006 MEDCOM briefed the Army Energy Strategy to its Huntsville Center of Expertise design managers, stressing the need to use LCCA, instead of least first cost, in the selection of new equipment.

WHS

No LCCA were implemented in FY 2007. LCCA has been implemented in prior years in the PENREN Project.

B. Retrofits and Capital Improvement Projects

DoA

Retrofits and Capital Improvement Projects which incorporate energy conservation are routinely implemented in Army projects when the LCCA indicates a valid return on investment. ECIP is a key component of the Army's energy management strategy. ECIP projects focus on energy, water, and cost savings; implementing renewable energy; and converting systems to cleaner energy sources. Utilities Privatization projects also result in energy conserving upgrades as a result of retrofit and capital improvements to the Army's utility systems.

DoAF

- Nellis AFB, Nevada, applied a ceramic paint coating to 24,000 square feet of decorative glass on facility in desert climate. New paint coating blocks more than 90 percent of solar radiation in blistering desert environment reducing solar heat gain.
- ACC instituted a "Pay Your Excess - Keep Your Savings" funding policy, which drove aggressive actions and resulted in $5.9 million in utility savings for reinvestment in energy initiatives such as: - Motion detectors - Photo cells for exterior lighting

o Metering

- Day-lighting
- High efficiency lighting
- Invested a total of $8,000 in Compact Fluorescent Lighting (CFL) at the Model Energy Bases in 2007. Purchased approximately 4,000 CFLs.
- Columbus AFB, Mississippi, installed Window Solar Film in their headquarters building.
- Little Rock AFB, Arkansas, used a technology-specific ESPC for the HVAC system retrofit with ground source heat pumps.
- The 611th Air Support Group at Elmendorf AFB, Alaska; replaced water mains and a waste heat loop at Eareckson Air Station, Alaska; replaced incandescent heaters with radiant heat in radar domes; motion sensors and timers were installed at two radar sites; also, installed waterless urinals at Barter Island, Alaska.
- Peterson AFB, Colorado, retrofitted one hangar with a solar wall and infrared heating.
- DeCA awarded lighting efficiency upgrades through group re-lamping and lighting retrofit projects for DeCA commissaries at Goodfellow AFB, TX, Fort Ord Community, CA, and McGuire AFB, NJ.

DLA

DLA completed six projects including lighting upgrades and HVAC control system upgrades.

NSA

Various energy savings projects were undertaken in FY 2007, which include: Installing variable speed drives on pumps and cooling towers, on-going roofing projects, re-lamping, building renovation projects, and Energy Management Control System (EMCS) replacement

WHS

The Pentagon is currently going through an extensive renovation of wedges 2 through 5. This renovation includes the replacement of all building systems using more energy efficient technology.

C. Use of Performance Contracts

DoA

ESPC is an important tool for financing energy efficiency measures that allow installations to improve their infrastructure and pay for the energy efficiency measures through the savings generated by the project over a 10 to 25 year time period. ESPC is a partnership with a private sector energy services company. Projects of note include the following.

Fort Jackson, South Carolina, awarded a $5.0 million ESPC task order at the end of FY 2007. The project includes EMCS, building recommissioning, thermal energy storage, substation upgrades, and central plant improvements. An additional $1.6 million in services is included in the task order. This task order was done under the DOE Super ESPC program.

Aberdeen Proving Ground, Maryland, awarded an ESPC contract for a $6.1 million steam system rehabilitation project. Implementation of advance controls was also included in the project.

U.S. Army Garrison Vicenza, Italy, awarded a $2.2 million ESPC contract for a 1.5 MW cogeneration project.

UESC, like ESPC, is an important tool for financing energy efficiency measures that allow installations to improve their infrastructure and pay for the energy efficiency measures through the savings generated by the project over time. The most notable difference between an UESC project and an ESPC project is that the UESC projects are financed and implemented through a utility company. Of importance to note for an UESC project is that the savings from the project are not guaranteed as is the case for an ESPC project. Notable FY 2007 UESC projects include the following.

- Fort Knox, Kentucky, awarded 5 UESC task orders. Three of the five included geothermal heat pumps, lighting retrofits, cool roofs, and steam boilers. The total investment value was $18.7 million.
- Fort Rucker, Alabama, awarded one task order with South Alabama Electric Co-op for an EMCS for $4.88 million.
- Fort Campbell, Kentucky, awarded three task orders for a total of $14.3 million. Two of the task orders were with Pennyrile RECC, and one was with the Tennessee Valley Authority. Both utilities serve Fort Campbell. The task orders were for EMCS and other electrical system improvements.

DoN

Based on past projects, DoN estimates it needs to invest $250 million annually in energy efficient equipment (financed + appropriated) in order to meet EPAct 2005 and EO 13423 energy reduction goals. UESC and ESPC are invaluable financial mechanisms to fund energy efficiency measures. NAVFAC, utilizing Utility, DOE, DoA, and DoN contracts, executes both contract vehicles and makes full use of appropriated project funds.

DoN awarded $84.6 million (first cost) financed energy projects that will provide 557,208 million Btu annual energy savings once constructed. Projects range from a $360,000 facility upgrade to a $32 million ground source heat pump and steam plant decommissioning project. Once the contract payments are completed, these projects will produce $63 million life cycle savings to DoN.

The following is a list of delivery orders awarded this fiscal year:

Activity Name	Project Title	Program
NSA Washington	Lighting Systems	ESPC
NAB Little Creek	Facility Energy Improvements	ESPC
NAS Oceana	Facility Energy Improvements	ESPC
NAS Jacksonville	EMCS / HVAC Controls	UESC
USNA Annapolis	Lighting Systems	UESC
USNA Annapolis	Lighting Systems	UESC
NAS Lemoore	Facility Energy Improvements	UESC
CNR Northwest	Facility Energy Improvements	UESC
CNR Northwest	Facility Energy Improvements	UESC
NAS Pensacola	Facility Energy Improvements	UESC
NAS Pensacola	Facility Energy Improvements	UESC
NAS Whidbey Island Steam and Condensate Systems		UESC
MCB Camp Pendleton Facility Energy Improvements		UESC
NAVFAC Southeast HVAC		UESC

The Air Force awarded three new ESPC and five new UESC task orders for FY 2007. These task orders include energy infrastructure upgrades and new equipment to help the installations reduce energy and water consumption and produce renewable energy. Examples include GSHP, boilers, lighting, motors, EMCS, and water reducing devices.

ESPC Table of Awarded projects

BASE	Award Date	Awarded To Total Contractor Investment	Contracting Agent
MacDill AFB, FL		$12,880 K	DOE
Hill AFB, UT	20 Oct 06	$4,564K	DOE
Hill AFB,UT	28 Sep 07	$7,693K	DOE

UESC Table of Awarded Projects

BASE	Award Date	Awarded To Total Contractor Investment	Contracting Agent
Minot AFB, ND	14 May 07	$12,761 K	AF
Offutt AFB, NE	28 Aug 07	$962 K	AF
Offutt AFB, NE	1 Mar 07	$999 K	AF
Shaw AFB, SC	6 Dec 06	$5,800 K	AF
Tinker AFB, OK	23 May 07	$895 K	AF

In FY 2007, the Department of Defense was allocated $55 million ECIP. The Air Force was successfully approved for 14 ECIP projects in FY 2007 for a total of $15,722,000. Projects ranged from as little as $310,000 for solar hot water heaters to as much as $2.8 million for extensive high efficiency landscaping projects.

Installation	State	Project	Cost ($K)	SIR*	Payback* (Years)
Nellis	NV	Facilities Efficiency	$820	4.81	2.68
Minot	ND	Repair HVAC/ELEC Community Center, Bldg 202	$687	1.8	7.65
Hill	UT	Condensate Return Polisher	$760	2.66	4.78
Beale	CA	Gray Water Irrigation Main Base	$1,560	2.54	7.49
Eielson	AK	Weatherize/Upgrade Various	$900	1.94	8.90
Kadena	JA	Install Water Saving Devices	$1,200	1.72	2.71
Ramstein	GE	Replace Airfield Ramp Lighting	$325	1.71	8.78
Davis-Monthan	AZ	Install Ground Source Heat Pumps in TLFs	$690	1.60	8.56
Tin City	AK	Wind Generator 250 kW	$1,650	1.54	7.49
Channel Islands	CA	Install GSHP, Wing HQ/Base Ops/CE	$1,750	1.42	9.81
Hurlburt Fld	FL	Replace Taxiway Lights with LED	$400	1.33	12.49
Nellis	NV	Install Water Efficient	$2,820	1.25	8.76

Installation	State	Project	Cost	SIR*	Payback
Fresno Yosemite	CA	Construct Photovoltaic Power System, Phase 2	$1,850	1.23	10.74
Laughlin AFB	TX	Dormitory Solar Hot Water	$310	1.48	9.50

* Weighted Average.

FY 2007 Energy Conservation Program (ECIP) Projects			
LOCATION	PROJECT DESCRIPTION	MBTU Savings	Cost Savings-$
Nellis	Facility Efficiency Improvements, Bldg 20	14,261	333,933
Minor	Repair HVAC/Elec (GSHP) Comm Ctr, Bldg 202	8,687	70,800
Nellis	Install Water Efficient Landscaping	0	75,715
Hill	Condensate Return Polisher, Various Facilities	19,011	100,489
Beale	GrayWater Irrigation Main Base	0	148,106
Eilson	Weatherize/Upgrade Various Facilities	7,287	89,517
Kadena	Install Water Saving Devices	29,264	263,607
Tin City	Wind Generator 250 kW	4,211	223,520
Hurlburt Fld	Replace Taxiway Lights with LED	131	2,252
Fresno Yosemite	Construct PV Power Generation System, PH II	1,948	90,797
	TOTAL	84,800	1,398,736

Defense Commissary Agency (DeCA)

DeCA is considering giving Huntsville District authority to make another attempt to pursue an ESPC. DeCA is also working with DOE Northeast Region for ESPC services pending completion of DOE partnering energy audits.

NGA

NGA's St. Louis Site contracted for an ESPC in 1999 and continues to realize savings from this effort. The payback period ends in FY 2012.

TMA

MEDCOM installations access all of the financing mechanisms available to them, including ESPC, UESC, utility company rebates, and appropriated funds. MEDCOM did not issue any new ESPC or UESC task orders in FY 2007. In FY2 007, MEDCOM received $750,000 in ECIP project funds for energy improvements at Tripler Army Medical Center, and projects $1.1 million in FY 2008 ECIP funding for projects at Womack AMC, Fort Bragg, NC, and Tripler Army Medical Center, HI. In August 2007, MEDCOM submitted two candidates totaling $1.7 million for FY 2009 funding.

WHS

The Energy Savings Performance Program was established between DOE and WHS in December 2006. This long term commitment allows WHS to utilize DOE's energy savings contractors to develop and implement creative ways to save energy. The program relates to one of the strategic goals for WHS and EO 13423 to reduce energy usage in Federal buildings. As a result Water Side Economizers at PH&RP and boiler installation at FOB2 were awarded. Future projects will include but are not limited to optimizing the PH&RP, energy awareness campaign, and infrastructures improvements. An ESPC partnership has been established with Honeywell to help meet the audit goals of EO 13423. The Pentagon will facilitate energy audits with its ESPC partner to identify energy saving strategies. Upon identifying and evaluating these strategies, the Pentagon will select the most practical strategies and authorize implementation.

D. Use of Energy Star® and Other Energy-Efficient Products

DoA

Use of ENERGY STAR and Other Energy-Efficient Products is an ever-increasing tool used by the Army as an energy savings strategy. ENERGY STAR is a program developed by the U.S. Environmental Protection Agency to promote energy efficiency in buildings. Installations as a practice consider ENERGY STAR products when purchasing new or replacement appliances and equipment. ENERGY STAR equipment must be included where applicable in all new Army construction.

DoN

The DoN eBusiness Office, Card Management Group, agreed to incorporate into the curriculum relevant information about federal buyers being directed to purchase products that are ENERGY STAR labeled or FEMP identified products. This training is provided by the Naval Supply Corps School and developed through the Card Management Group. GSA is a regular instructor at the

Navy in-house energy manager's course and has partnered with the DOE and the Environmental Protection Agency to offer the federal community a broad range of energy efficient products. They have identified energy efficient products in catalogs and on the GSA Advantage, an online shopping and ordering system that provides access to thousands of contractors and millions of products and services.

A survey was conducted of energy efficient products policy and products use. Of 92 installations responding, the following table indicates strong use of available information and incorporation of energy efficient products.

Energy Efficient Products Issue	# of Installations responding
Recommendations Distributed and Use is Encouraged	76
Advertised During Awareness Activities	68
Instructions Address FEMP Recommendations	65
Designers and Equipment Specifiers use FEMP product recommendations	77
Procurement Officials use FEMP Recommendations	69
Nothing Formal is Being done to Implement FEMP Recommendations	9

ENERGY STAR performance criteria are included in acquisition requirements for systems and appliances in privatized family housing units. Energy efficient operations are included in some Base Operating Support Contracts.

DoAF

The Air Force continues to pursue a policy that all purchases of computers, printers and copiers will be specified as ENERGY STAR compliant as stated in the EPAct 2005. Design specifications for new and retrofitted equipment are reviewed to ensure they are in the upper 25 percent or ENERGY STAR compliant as stated in Unified Facility Criteria 03-400-01, Design: Energy Conservation. Examples include:

- Minot AFB, South Dakota, installed ENERGY STAR HVAC, lighting, windows, appliances, and insulation in 1,700 new military family housing units saving $1 million per year.
- Nellis AFB, Nevada, replaced refrigerators, stoves, and other appliances with ENERGY STAR equipment in 1,178 military family housing units.

- The 611th Air Support Group, Elmendorf AFB, Alaska, uses ENERGY STAR motors in several locations for the HVAC system.
- Kunsan Air Base, Korea: ENERGY STAR appliances (123 washers, 73 refrigerators, and 360 dehumidifiers) purchased in FY 2007 will save the base at least $17,000 in electricity annually.

DCMA

DCMA uses ENERGY STAR products such as computers, printers, etc.

DeCA

ENERGY STAR products continued to receive utmost consideration when developing specifications and issuing acquisitions for energy using products. Information technology hardware and computer and copying equipment are acquired under the ENERGY STAR program using GSA schedules and either government-wide or service contracts. DeCA continues to review and pursue opportunities to utilize electronic communication and data transfer, which conserves the use of such things as paper, postage, and personal resources. Additionally, ENERGY STAR compliance is a requirement for vendors to participate. One hundred percent of the Program Management Directorate acquisitions are energy efficient. DeCA published a hardware standards document to assure 100 percent energy efficient hardware purchases. DeCA design criteria requires premium efficiency fan motors for HVAC systems, electronically commuted fan motors on the refrigeration display cases, and T- 8 fluorescent light fixtures with electronic ballasts in display cases and in new and renovated facilities.

DeCA's Contracting Directorate procures energy efficient products, such as paper grocery bags made up of a minimum of 5 percent pre-consumer or post-consumer recycled products. New or replacement balers are purchased for our commissaries in consideration of efficient disposal of cardboard products. In FY 2006, the DeCA Government Purchase Card (GPC) Program began requiring all new GPC program participants complete training on the ENERGY STAR Program prior to being issued an account. ENERGY STAR training is part of the Mandatory Services and Products Training Module, and is provided via Power Point slides posted in the DeCA Public Folders portion of Microsoft Outlook. This users' training includes a description of the requirement, how DeCA must comply, and provides an introduction to tools to find ENERGY STAR compliant products. The level of training is for the purchase card user and the supervisor. ENERGY STAR training is also provided in a classroom setting during 3-day GPC Hands-on Training. Hands-on training is approximately 8 times per year. In

FY 2007, all current GPC participants were required to complete the ENERGY STAR regardless of how long they have been a participant. Benefits are reduced energy costs as more energy efficient products are purchased, improved efficiency in purchasing ENERGY STAR Products, and better understanding of Agency energy usage.

Higher level Affirmative (Green) Procurement Training sponsored by the Air Force has been taken by: the Chief, Contract Policy and Systems Division, contracting's environmental point of contract, as well as by the GPC Agency Program Coordinator to further the training and education of energy management policies at the mid and upper levels of DeCA management.

DFAS

DFAS promotes all sites to use Energy-Efficient Products when replacing items.

DIA's Directorate for Mission Services procures energy efficient products. DIA plans to procure an industrial shredding and briquetting system for disposal and recycling of its classified waste.

DLA

DLA uses USACE and NAVFAC for design of new construction and renovations. DLA also uses ENERGY STAR products such as computers, printers, etc.

NGA

NGA annually procures approximately 11,000 personal computers and monitors with ENERGY STAR enabled features.

NSA

NSA utilizes ENERGY STAR, and energy-efficient products for all new installation, renovation, and operations and maintenance projects. Energy efficiency language is incorporated in all new construction and renovation project specifications.

WHS

The Pentagon encourages the procurement of ENERGY STAR office equipment. In FY 2007, ENERGY STAR computers, copiers, and printers were purchased by WHS. Energy efficient light bulbs were purchased and average about 1000 lights per year.

E. Sustainable Building Design and High-Performance Buildings

DoA

The Army has embraced this concept and has identified projects since FY 2002 as Army sustainable building design showcase facilities. USACE has incorporated sustainability principles into its design and military construction transformation process.

- In 2006 the Assistant Secretary of the Army (Installations and Environment) issued a memorandum on SDD Policy to update the Army strategy for integrating the principles and practices of sustainability on our installations as we minimize the impacts and total ownership costs of Army systems, material, facilities, and operations. Accordingly, the Army will transition from the Sustainable Project Rating Tool to the USGBC LEED rating system effective with the FY 2008 MILCON program.
- Of the 384 new building (or major renovation) design/construction projects in FY 2007, 301 projects have been reported by installations as can or will be certified under LEED.

DoN

NAVFAC Instruction 9830.1, Sustainable Development Policy, June 9, 2003, implements sustainable development principles and strategies to reduce the total cost of ownership of facilities. The policy requires the use of the USGBC LEED System and that all new construction projects meet the LEED Certified level unless conditions exist that limit the pursuit and accomplishments of the LEED credits necessary for achieving the LEED Certified level. The ASN(I&E) Memorandum, August 4, 2006, directs Navy and Marine Corps Commanders to take steps to plan, program and budget to meet the requirements in EPAct 2005, the Federal Leadership in High Performance and Sustainable

Buildings Memorandum of Understanding and at least LEED Silver level rating performance in new and replacement buildings. NAVFAC is developing interim guidance to implement the ASN memorandum. NAVFAC Instruction 9830.1 will be revised to incorporate the requirements of the ASN memorandum and other necessary sustainable program updates such as programming, metrics, data collection, installation planning and LEED certification.

Currently, DoN has the following LEED certified buildings: 1Gold, 1 Silver and 5 certified buildings. Additionally, DoN has self-certified 4 additional Silver buildings and 30 certified buildings.

DoAF

Air Force SDD Policy was signed on July 31, 2007, it updates and expands existing policy (December 19, 2001, Sustainable Development Policy) and reinforces the importance of sustainable development concepts in the planning, design, and construction process, and operation of facilities and infrastructure. Beginning in FY 2009, 100 percent of each major command's MILCON vertical construction projects, with climate control, shall be designed so that it is capable of achieving LEED Silver certification. This is not an option; sustainable features can not be eliminated to save scope or cut cost. To accomplish this goal, the Air Force will document SDD, EpAct 2005, and EO 13423 costs on the DD Form 1391, with a separate line item under primary facility costs identified as "SDD & EPAct05", beginning with the FY 2009 MILCON program.

The Air Force has had over a dozen projects capable of achieving certification; and have proactively stepped out to formally certify (ahead of the July 31, 2007, policy memo mandates) the 11 projects listed below.

- Four projects have been formally certified with the USGBC: o FY 1999 Barksdale AFB, Louisiana, Physical Fitness Center o FY 2002 Edwards AFB, California, Consolidated Support Facility o FY 2003 Barksdale AFB, Louisiana, Dormitory

o FY 2004 Seymour-Johnson AFB, North Carolina, Fire Station

- Seven projects have registered with USGBC and are awaiting certification: o FY 2005 Shaw AFB, South Carolina, Library

o FY 2005 Barksdale AFB, Louisiana, Dormitory
o FY 2006 Offutt AFB, Nebraska, Air Force Weather Agency Headquarters o FY 2007 Shaw AFB, South Carolina, Dormitory
o FY 2007 Langley AFB, Virginia, Dormitory
o FY 2007 Langley AFB, Virginia, Distributed Common Ground Station o FY 2008 Tyndall AFB, Florida, Fitness Center

DCMA

During FY 2007, DCMA continued planning on one new building project for which sustainable building design were incorporated into the siting, design and construction.

DeCA

DeCA's Design Criteria has included sustainable principles for several years. The Criteria emphasizes use of life-cycle costs, pollution prevention, and other environmental and energy costs associated with the construction and life cycle operation of the facility. Detailed requirements are incorporated for items such as energy efficient lighting, dual-path HVAC, premium efficiency fan motors, refrigeration monitoring and control systems, lighting controls, and roof membrane materials. DeCA design criteria are reviewed and updated annually. Standard design principles used are as follows:

- Storm Water Management
- During site selections, sensitivity to habitat of endangered species and wetlands, preference to urban areas with existing infrastructure
- Advocate alternative transportation by providing bicycle racks and size automobile parking to meet, but not exceed what is needed
- Maximize water efficiency by using automatic water controls for restroom fixtures
- Implementation of a building commissioning plan
- Maximize use of energy saving techniques to include implementing heat reclaim from the refrigeration systems to provide space heating and water heating, using refrigeration compressor systems comprised of several compressor sizes to ensure that the most efficient combination of compressors are running at any one time to meet the load, implementing a RMCS for control of the Refrigeration System and HVAC System to ensure efficient operation, maximized use of more efficient glass door refrigerated display cases rather than open cases, maximized use of the most energy efficient lighting systems implementing the most advanced lighting equipment available, use of lighting occupancy sensors to ensure lighting is off when rooms are not in use, use of a dual path HVAC unit for the sales area which allows for a more efficient method of cooling, maximum use of wall and roof insulation, implementation of the most energy efficient doors and windows, and maximum use of utility metering
- Chlorofluorocarbons reduction in HVAC and refrigeration equipment

- Use of fire suppression systems that use no hydro-chlorofluorocarbons or halon
- Use of cardboard balers
- Maximum use of existing building structures, rather than replacing, through our major Add/Alt projects
- During building demolitions recyclable materials are saved
- Use of materials with recycled content
- Meet indoor air quality standards.

All new construction and design projects initiated in FY 2007 used sustainable development and design principles as a standard for DIAC construction. Standard design principles used are as follows:

- Storm Water Management
- During site selections, sensitivity to habitat of endangered species and wetlands, preference to urban areas with existing infrastructure
- Advocate alternative transportation by providing bicycle racks and size automobile parking to meet, but not exceed, what is needed
- Maximize water efficiency by using automatic water controls for restroom fixtures
- Implementation of a building commissioning plan
- Maximize use of energy saving techniques to include maximum use of the most energy efficient lighting systems, maximum use of lighting occupancy sensors and maximum use of utility metering
- Chlorofluorocarbons reduction in HVAC and refrigeration equipment
- Use of fire suppression systems that use no hydro-chlorofluorocarbons or halon
- Maximum use of existing building structures, rather than replacing, through our major add/alt projects
- During building demolitions recyclable materials are saved
- Use of materials with recycled content
- Meet indoor air quality standards

DIA is committed to sustainable development and incorporating sustainable principles in all major design and construction projects.

DLA

All of DLA design and construction of new facilities are done by the USACE or NAVFAC, which use Sustainable Building Design.

NSA

NSA did not construct any new facilities during FY 2007. However, NSA uses LEED Silver criteria in our engineering designs, and will be pursuing certification on selected projects.

TMA

MEDCOM encourages sustainable building design. All new construction meets the LEED "Silver" standard. Similarly, major renovations meet LEED "Silver" standards for the specific work being accomplished. To this end, MEDCOM reviewed the LEED standards to identify the areas that our renovation work could most effectively impact. Facility managers were then briefed on the standards and suggested areas of concentration during the Command's energy conference held in July 2007.

WHS

PENREN incorporates sustainability requirements and goals in each design-build RFP. Currently, the Pentagon wedges are being renovated and where feasible has incorporated sustainable design and construction practices. As a way to benchmark these sustainable measures, PENREN is using the USGBC LEED as the primary metric for determining sustainability success in buildings. The wedges and other structures built in association with the Pentagon have chosen to submit for certification under the LEED for New Construction rating system. The following projects are ongoing: 1) Wedge 3 started in FY 2005 and was completed in October 2007. The LEED submission is currently being finalized and the project anticipates receiving a "Certified" rating. 3) Wedge 4 construction has started and like Wedge 3, the building's design has required LEED-New Construction certification. 4) The Pentagon Memorial started construction in late FY06. Because it is an outdoor facility, LEED certification is unattainable. However, there will be efforts to identify LEED certification points within the design. 5) The Pentagon Library and Conference Center was awarded in July 2004. However, most of the design was accomplished in FY 2005. This design-build contract included a requirement that a "Certified" rating be earned under LEED for New Construction. The LEED submission is currently under review by the USGBC and the project anticipates receiving a "Silver" rating. 6) Wedge 2 was completed in September 2006 and has received a "Certified" rating from USGBC.

The ENERGY STAR Target Finder has evaluated the Pentagon Library and Conference Center design as in the top 12 percentile in energy efficiency.

F. Energy Efficiency/Sustainable Design in Lease Provisions

DoA

The Army emphasizes that energy and water conservation be included in all facility leases and requires that these leased facilities meet energy and water goals. The intent is to have the landlord make appropriate investments in energy efficiency, which can be amortized in the lease, provided the new total cost (energy costs plus lease cost) does not exceed total costs without improvements. Leases should amortize the investments over the economic life of the improvements. Build-to-lease solicitations for Army facilities contain criteria encouraging sustainable design and development, energy efficiency, and verification of building performance. The Army has no activity to report for Energy Efficiency in Lease Provisions.

DoN

Leased space must also comply with the energy and water efficiency requirements of the EPAct 1992. The Navy must rely upon GSA to ensure the above provisions are included in buildings that they lease for the Navy. The Department of Navy currently leases 57.5 million square feet of building facilities.

Air Force

The Air Force evaluates all leased properties for location, cost/square foot, sustainability and energy efficiency. All these factors are reviewed before accepting a lease. Examples include:

- 611th Air Support Group Elmendorf AFB, Alaska – Tenant lease provisions state that tenants are expected to practice energy conservation in their leased facilities.
- Eielson AFB, Alaska, has 1,028,875 square feet of leased housing facilities built in 1995, and 570,000 square feet built in 1985 and 1986. This housing meets current energy efficiency standards.

- CCAFS, Florida: All Support Agreements between CCAFS and tenants include a paragraph stating that tenants need to follow Air Force energy conservation guidelines (EPAct 2005 and EO 13423).

The Air Force actively pursued public-private partnerships to increase our energy supply options. The previously mentioned solar PV array at Nellis AFB, Nevada, is a prime example, where the Air Force leased 140 acres to a private entity who built and is operating the largest array in the Americas. Building on the momentum of that project, the Air Force Real Property Agency (AFRPA) hosted an energy-focused enhance-use lease Industry Forum in Ontario, CA, September 25-27, 2007, that brought together over 170 representatives from private industry, Federal and State Agencies, and the Air Force. At this event, the AFRPA discussed the Air Force's energy strategy and highlighted potential opportunities for projects capable of generating energy at AFBs throughout the country, including potential commercial grade utility scale solar energy projects at Edwards AFB, CA, Kirtland AFB, NM, and Luke AFB, AZ.

DCMA

At facilities where DCMA is a tenant, DCMA activities have been instructed to follow the host's energy program.

DeCA

DeCA uses GSA as a leasing agent for its few leased facilities. Lease-back facilities include the Harrison Village Commissary, IN; DeCA West office facilities, Sacramento, CA; and the McClellan Commissary, Sacramento, CA, were negotiated or initiated in FY 2007. Consideration was given on all procurements to include high efficiency systems for HVAC and water use. Devices that meet or reduce current flow rates for water use are used. The regional office contracted for the space and requested that the building owner incorporate current commercial energy efficient design with set back thermostats and state-of- the-art equipment Seasonal Energy Efficiency Rating ratings. The plumbing fixtures in the office are all of the low-flow type and meet commercial plumbing criteria. The energy and utility costs are included in the lease agreement so the rent includes these costs.

This makes it beneficial to the owner to use the most cost effective energy-using equipment to maximize their energy savings. No new DeCA West office facility leases were negotiated or started in FY 2006. Regional office facilities and the McClellan Commissary are being occupied under a lease-back through the Air Force Real Property Agency (AFRPA). These facilities were occupied in an "as-is" condition, and energy efficiency measures (i.e., occupancy sensors, low-flow type fixtures, energy efficient lighting) have been installed at DeCA's expense to reduce energy costs.

DeCA Europe continues to work with the base installation command or the U.S. Army garrisons to negotiate with the leased property owners. We ensure that the existing facilities meet energy efficiency and water conservation requirements, by requiring all plumbing fixtures to be water saver types, with energy efficient lighting, and insulated walls and roof.

DIA

DIA uses the GSA / WHS as a leasing agent for its few leased facilities. DIA uses energy-savings provisions contained in GSA / WHS leasing regulations.

NGA

NGA occupied a third leased building on the Reston site in late FY 2006. The use of energy efficient technologies was incorporated into lease provisions, resulting in the installation of energy reducing devices including ENERGY STAR appliances, heat exchangers, low flush toilets, electronic water valve controls and variable speed fans. As the building nears the end of its first full year of operation, NGA continues to work with the leasing agent to insure building systems are maintained in optimal working condition in order to gain the full benefit from installed energy efficient technologies.

NSA

NSA has several leases where energy efficient methods and products have been used for renovations and repairs. Employees occupying or working in these spaces follow all existing agency energy policies. When entering into a lease for unfinished space that NSA will fit-up, we employ the same design standards used for our new buildings and renovation projects.

WHS

Most of the space leased for WHS uses GSA leases. There are however, three DFD leases which use the same GSA lease provisions. The typical Solicitation for Offerors informs the lessors to comply with the government's energy conservation guidelines. The Lease Facilities Division (LFD) supports that in their day-to-day lease administration. When the LFD adds equipment for special requirements, they try to influence the lessors to use energy efficient applications.

G. Industrial Facility Efficiency Improvements

DoA

Army installations are actively pursuing energy efficiency of industrial facilities. Ongoing measures highlighting the Army efforts are discussed.

- Over 60 percent of the energy used by Anniston Army Depot, Alabama, is for industrial processes. The site-wide Federal Energy Decision Screening assessment conducted by the DOE Pacific Northwest National Laboratory (PNNL) identified $5.4 million in potential retrofit projects in the buildings and facilities. The annual savings from these projects is identified at $700,000 per year. Anniston made use of Army Working Capital Funds to replace steam heaters in a number of large production buildings with gas-fired infrared heaters and to replace old lighting with new T5 lighting with controls. Anniston is also actively pursuing ECIP funds for projects.
- Bluegrass Army Deport, Kentucky, also has substantial industrial processes amounting to approximately 23 percent of their total energy use. A PNNL assessment identified many opportunities to improve energy efficiency. Blue Grass Army Depot ultimately submitted a FY 2007 ECIP project as a result of the assessment.
- Aberdeen Proving Ground, Maryland, has focused their efforts on day-lighting and solar tubes to offset the enormous electrical costs associated with high-bay lighting.
- Watervliet, New York completed numerous FY 2007 projects resulting in considerable improvement to the arsenal's building and equipment energy efficiency. Work included wash water recycling technology as well as off-peak power use.
- The Joint Systems Manufacturing Center, Lima, Ohio, has installed energy efficient lighting fixtures throughout the main production areas. Efforts are ongoing to reduce energy consumption through reduced boiler operations during summer months and reduction of steam levels for weekends and off-shifts. Several walk doors have been replaced throughout the facility to improve sealing against the outside environment. Ventilation control modules are being replaced or updated for a restricted access building, which has no doors or windows that can be opened for ventilation. New control modules for the existing EMCS will allow for more efficient operation of the restricted-access building.
- Rock Island Arsenal, Illinois, has undertaken a two-year program to eliminate all steam absorption chillers. The project culminated this

summer with the replacement of approximately half of the Garrison's chiller capacity. The installation also spent $8 million to install state-of-the-art controls, high efficiency burners, oxygen trim controls, coal-feeding improvements to decrease combustion losses, and ductwork repairs to better control flue gas and eliminate leaks.

DoN

- Puget Sound Naval Shipyard, WA, replaced nearly 170 motors ranging in size from 3 to 200 HP with premium efficiency motors.
- Naval Base Point Loma, CA, retrofit two chiller compressors with oil-free variable speed compressors at the Old Town Complex Bldg. 2.
- Naval Base Coronado, CA, audited and made adjustments to several industrial facilities as part of their Building Tune-Up Program.
- Naval Facilities Engineering Command Southwest, CA, carried out a major upgrade of South Cummings substation at Naval Base San Diego to greatly improve reliability of the power supply to the fleet while reducing demand charges.
- Naval Air Station Oceana, VA, participated in a Navy Techval demonstration of thermal destratifiers in Hangar 145. Thermal destratifiers save energy by moving and mixing the air throughout the hangar area, preventing pockets of cold and hot air.
- Naval Facilities Engineering Command Hawaii completed water, electrical and waste water treatment plant projects, increasing system efficiencies.

DoAF

- Columbus AFB, Mississippi, is coordinating assessment of commercial air service standards air compressors and infrastructure to increase efficiency.
- Eielsen AFB, Alaska: Projects completed include a warehouse retrofit with T5HO lights and motion sensors, head bolt outlet intelligent parking lot controller and meter installation, wooden door replacement at Aircraft Ground Equipment maintenance facility and reside/insulation of several buildings.
- Innovative Airfield Lighting Project.; Pope AFB, NC. Pope AFB has 6 ramps. The base has 41 poles used to illuminate the ramps. The lights consume an estimated $218,000 worth of electricity each year, or over 7 percent of Pope's total usage. Add $100,000 of maintenance expenses –

replacing bulbs, ballasts, and rental equipment – gives a total annual cost of $318,000. Before the control project the ramp lights were photocell activated, burning each night from dusk to dawn. Pope Operations did not turn off the lights when aircraft were not present, rather the switch was left in "auto" controlled by the photo cell. No power to the pole shut "off" the red obstruction light on each pole. No obstruction lights were unacceptable. The base developed a radio frequency control system with an encrypted signal for each pole lighting control system so that individual pole aircraft ramp lights could be turned "on" or "off" without shutting "off" the red obstruction lights. This project was fully operational in Aug 2007. The project cost $313,000 and has a simple payback of 2 years. The control system reduces ramp lighting energy consumption by 51 percent (approx $100,000) and associated lamp replacement costs by $46,000 per year. Pope AFB used a modification to an existing UESC to fund the contract.

DeCA

DeCA installs dual-path air conditioning to control commissary store humidity as an alternative to natural gas or propane fired desiccant dehumidification systems. DeCA has increased the use of heat-pipe technology for dehumidification and heat reclaim. Domestic hot water and HVAC heat reclaim systems are standard in most large commissary systems. A QSR at each commissary (store) monitors refrigeration and HVAC maintenance contract performance. DeCA conducts remote diagnostic monitoring of RMCS at approximately 191 individual stores to ensure that refrigeration and lighting systems are being operated and maintained at their design specification. Discrepancies are forwarded to our maintenance contractors on a daily basis for correction.

Lighting controls were monitored and adjusted by this same method in FY 2007. This surveillance resulted in improved contractor maintenance and improved equipment operation and less energy consumed. Web-based energy monitoring control systems are being evaluated for DeCA-wide use. Energy efficient lighting and occupancy sensors are installed in all new construction and renovation projects. Computers are turned off at night and on weekends. Automatic lighting systems are installed in most refrigerated cases to turn off lights after business hours. Plastic or metal swing air curtain doors are installed where feasible. A lighting study is currently underway to identify new lighting equipment and strategies in order to reduce the energy consumption (watts per square foot) in order for commissary facilities to meet the requirements of the

ASHRAE Standard 90.1, Energy Standard for Buildings Except Low-Rise Residential Buildings. The proposed design is expected to reduce the life-cycle lighting costs by 12 to 30 percent.

Regional Utilities Energy Efficiency Task Forces (UTF) are active in DeCA East, West, and Europe.

This integrated approach to solving energy efficiency and billing problems has been successful in identifying actual and potential energy savings with the objective of reducing total operating costs. DeCA UTFs evaluated high energy use facilities to determine causes and what should be done to correct problems. Each store's unit cost for utilities is monitored; all that are indicated as exceeding other stores in the same sales band are investigated for causes and recommended for correction. Identification of utility costs as a category of the unit cost program has intensified the store management's interest in the cost of water, electricity, and other utilities that are billed to the store.

Stores, where restroom facilities have been upgraded, have been fitted with the low-flow-type fixtures and, where applicable, the sensor activated faucets and flush valves. The process is ongoing. DeCA awarded lighting efficiency upgrades through group re-lamping and lighting retrofit projects for DeCA commissaries at Goodfellow AFB, TX; Fort Ord Community, CA; and McGuire AFB, NJ.

DeCA should receive a Sacramento Municipal Utility District energy rebate in the amount of $9,833. Store sales area designs include energy efficient ballasts and lamps, state-of-the-art HVAC systems, and lighting circuited and controlled by time-of-day schedules through the RMCS. DeCA Europe continues to identify and replace refrigeration systems with newer more energy efficient refrigeration systems. Several small projects were initiated in FY 2007 to replace portions of aging refrigeration systems. During FY 2007, significant energy efficient replacement refrigeration systems, projects were completed in DeCA East commissaries at the Fort McCoy Commissary, Mayport Commissary, and Seymour Johnson Commissary. Several roof projects were completed in FY 2007, including the replacement of the deteriorated insulation. Installation of a light reflective roof surface improved the energy efficiency of the entire roof system.

Replacement of 21 inefficient ammonia evaporators in the Cold Storage facility was completed in FY 2007. This will help reduce the energy consumption due to old and leaking evaporators.

A replacement project for coffin-type frozen food display cases with Glass-Door up-right display case was completed at Vicenza Commissary to reduce the energy consumption of the low temperature compressor racks.

During FY 2007, two new commissary facilities were constructed within DeCA West (NS San Diego, CA, and Peterson AFB, CO). New equipment and a modern facility not only enhanced the shopping experience of our patrons, but also dramatically improved the energy efficiency and cost saving for all utilities.

DeCA West completed nine lighting upgrade projects and four others have been awarded to provide either the entire facility or large portions of the facility with a more energy efficient lighting system.

DeCA West awarded one roof project in FY 2007, which will incorporate a reflective coating that has proven to dramatically improve energy efficiency. The project includes replacement of deteriorated insulation.

DeCA West replaced a deteriorated and inefficient RMCS.

To ensure DeCA West has the most energy efficient system, all new and/or renovated facilities have a combined/integrated cooling, heating, incorporated as a standard practice in each design.

H. Distributed Generation, Including Combined Cooling, Heating, and Power Systems

DoA

The Army policy calls for high efficiency equipment in the operation of central heating and cooling systems where large quantities of energy are used. The Army budgeted FY08 funds to begin a centrally-funded program to modernize aging central heating systems. Some installations continue to add Digital Direct Controls and other automated controls to better monitor and control energy. Installations also use Operation & Maintenance funds to implement energy saving projects such as: upgrading boilers and distribution systems, improving high efficiency pumps and motors, and updating control system. Army regions and installations, along with USACE, evaluate the use of high-efficiency energy systems for new construction and major retrofit projects and incorporate these systems where cost-effective.

- Construction started in FY 2007 on an ECIP project at Fort Stewart, Georgia, to install a Hydrogen Fuel Plant. This project will produce on site electric power and will provide a 200 kW generator fueled by hydrogen gas recovered from waste gas from the waste water treatment plant's anaerobic digester. Additionally, the hot water generated by the fuel cell will be used to meet part of the installation's hot water requirement.

- The Presidio of Monterey, California, installed a 60 kW Capstone micro turbine at the Child Development Center that provides 60 percent of the electrical and heat load. Reuse of the exhaust by-product meets most of the space heating requirements in the building. Future plans are to add production capability for sufficient hot water to meet all thermal requirements.
- Fort Hood, Texas, implemented a year-long fuel cell demonstration project also serving as a backup for emergency generator power at a communication facility in FY 2007. The final report of the fuel cells performance has yet to be published.

DoN

DoN utilizes distributed generation primarily for island installations, remote applications, and in technology demonstrations.

Wind farms and PV systems continue to provide power at numerous installations. New renewable power systems were made operational at: Naval Base Ventura County, CA, and Naval Support Detachment Monterey, CA.

DoN continues to validate the performance of fuel cells.

In other fuel cell developments, Proton Exchange Memebrane fuel cell power plants, powered by an onboard supply of methanol, and incorporating a new design non-fluorinated Membrane Electrode Assembly, were brought on line at Naval Air Station, Pearl Harbor, HI.

Marine Corps Base Camp Pendleton, CA completed commissioning tests for a 500 kW Molten Carbonate Fuel Cell and when complete the total capacity of this fuel cell installation will be 750 kW. An additional 300 kW Molten Carbonate fuel cell installation is planned for Pacific Missile Range Facility, HI. These installations will validate performance and cost in a combined heat and power application.

DoN is validating the performance of energy technologies such as cool roofs, heat pipes, air conditioning compressors with integrated variable speed drives, air conditioning duct sealants, high efficiency air conditioning systems, scotopic lighting and destratification fans. The results of the demonstrations will be used to guide installations on the life cycle cost benefits of using these technologies to reduce electrical loads. Other electrical load reduction measures taken in FY 2007 are shown below:

- All installations were directed to operate spaces at no cooler than 78 F during cooling season and no warmer than 68 F during heating season.

- Naval Undersea Warfare Center, Division, Keyport, WA, conducted several after-hours energy walk-throughs in numerous facilities to identify lights and office equipment that may have been left on. Keyport monitored electric meter data and made recommendations to internal customers to consider production schedule adjustments to avoid high load time periods.
- NAVFAC Southwest, CA, created a Renewable Energy Plan to develop wind, photovoltaic and other renewable resources and started to execute near-term actions. They provided material support to the Navy's first Shipboard Energy Management Pilot Project at Naval Base San Diego and Naval Base Coronado to help reduce fleet energy and water usage and utility costs. They continued a Building Tune-Up Program to reduce demand in energy-intensive facilities, participated in a utility-provided Air Conditioning Tune-up Program, and conducted actual load sheds at bases. They continued to expand EMCS DDC systems throughout San Diego naval facilities.
- Fleet Readiness Center Southwest, CA, implemented Airspeed Lean/Six Sigma production projects throughout the facility.
- Naval Air Facility El Centro, CA shut down five unoccupied bachelor quarters by setting temperature set points at 90 degrees or shutting off the HVAC units. All appliances were unplugged while the buildings remained empty.
- Naval Base San Diego, CA, used data from time of use meters on each building to help identify load reduction opportunities.
- Naval Base Ventura County, CA, participated in the utility company demand bidding program to reduce peak period loads, and conducted quarterly load-shedding drills to reduce peak loads and prepare for electrical emergencies.
- Naval Base Coronado, CA, re-commissioned DDC EMCS hangar bay lighting controls throughout the base.
- Naval Support Detachment Monterey, CA, reduced lighting campus wide by removing 50 percent of fixture lamps in common areas. The Command Duty Officer monitored, recorded and reported energy inspections on-line on a nightly basis. Monterey used their EMCS Service Contract to its maximum extent to fine tune all equipment to achieve the greatest savings.
- Naval Weapons Station Seal Beach, CA, raised chilled water temperatures of building cooling systems.

- Marine Corps Air Station Camp Pendleton, CA, replaced in-ground incandescent taxiway lamps with LED fixtures.
- Marine Corps Base Camp.Pendleton, CA, expanded their EMCS system.
- Naval Support Activity Norfolk, VA, implemented lighting improvements.
- Naval Air Station Meridian, MS, replaced HVAC systems in eight buildings to reduce electric consumption 209 MWh per year.
- Marine Corps Base Camp Lejeune, NC, retrofit existing lighting with higher efficiency ballasts and lamps.
- Marine Corps Recruit Depot Parris Island, SC, upgraded HVAC systems, motors, and lighting base-wide.
- Naval Station Pearl Harbor, HI, trained over 870 building energy monitors and funded $1.9 million in energy efficient lighting and solar film energy saving projects.
- Fleet Activities, Sasebo, Japan, tracked monthly consumption metrics and provided them to the largest consumers, including the homeported ships.
- Naval Station Rota, Spain, secured utilities to vacated buildings.
- Naval Support Activity Souda Bay, Crete, Greece, increased energy awareness for lighting and small appliances through Plan of the Week messages.
- Naval Region Contracting Center Singapore reduced the wattage and quantity of perimeter lighting, and reduced passageway lighting in the building.

DoAF

- Goodfellow AFB, Texas, has an aggressive peak load shaving plan, preset actions to reduce electrical load as temperatures increased.
- Lackland AFB, Texas, and Randolph AFB, Texas, uses EMCS controlled load shedding program.
- Sheppard AFB, Texas, received the FEMP award for "No Cost Load Shedding Program" and an annual load reduction program that is implemented during air conditioning season where 50 percent of the lighting is turned off.
- Vance AFB, Oklahoma, uses Thermal Energy Storage and an EMCS controlled load shedding program.
- Elmendorf AFB, Alaska, Distributed Generation, including combined cooling, heating, and power systems: the base's former central steam

plant and distribution system was eliminated and replaced with a 280-boiler distributed heating system in 165 facilities.

- The USAFA utilizes biomass energy generated in the digesters at its on-site Waste Water Treatment Plant to produce hot water required by the waste treatment process. This represents a highly efficient and self-sustaining application that the Academy is further developing by installing a small, combined heat and power MicroTurbine generation system.

Energy Collaboration with Allies and Coalition Partners. The Air Force began military to military collaboration with our allies and coalition partners on facilities and infrastructure energy in 2007.

For example, SAF/IE and members of his staff met with the senior leadership and staffs of the United Kingdom's Royal Air Force and French Air Force. They discussed common areas of interest for energy conservation, sustainable design and alternative energy supplies. They have mapped out an aggressive engagement plan to share best practices, including aggressive energy management plans and public-private partnerships to generate energy on underutilized military property. The Air Force plans to expand this engagement strategy to other allies world-wide in 2008.

DeCA

If the installation cannot provide dependable power to the commissary, DeCA's programming and design team reviews what other sources are available to supplement the power for the commissary building. DeCA typically uses generators for backup power of Point of Sale and emergency lighting systems only.

Geothermal heat pump systems are considered as design alternatives when cost effective. PV systems are surveyed and a PV project at Los Angeles AFB Commissary was funded in FY 2006 for award in FY 2007. New refrigeration systems utilize electronic controls, heat reclaim, and refrigeration compressor "floating head" to reduce energy usage. Utilities are normally provided by host installations to DeCA. Designs include heat reclaim on refrigeration for HVAC and domestic hot water. All new and/or renovated facilities have a combined/integrated cooling, heating, and refrigeration monitoring system incorporated as a standard design. Combined and integrated cooling, heating, and refrigeration systems are a standard design concept for installation of new or replacement refrigeration systems in the commissaries.

Significant energy efficient replacement refrigeration system projects were completed in FY 2006 at the DeCA East Maxwell AFB Commissary and Gulfport Commissary. Similar projects that incorporate energy efficient refrigeration systems are ongoing and were completed in FY 2007 at Mayport Commissary, Key West Commissary, and Seymour Johnson Commissary.

DIA

DIA has on-site generation capability for 100 percent backup power. Peak shaving is not performed due to the lack of a considerable demand charge for electricity.

EMCS to ensure energy efficiency of the facility's overall operation are installed in all new/renovated facilities as standard design criteria. Standard motor starters were replaced with variable speed drives for all variable air flow systems in the DIAC facility, in conjunction with air handling unit replacements. The savings are considerable as the motor uses far less power when commanded to reduce speed as opposed to letting fans simply ride their characteristic pressure-airflow curve at 100 percent speed while adjusting inlet guide vane position.

ENERGY STAR power-down features are activated on most electronic equipment. Personal appliances such as coffee pots and radios are secured at the end of the workday. Personal electric space-heating devices and desk fans are prohibited.

DIA's standard practice is to install motion sensors and separate lighting circuits to allow turning off unneeded lights.

TMA

No MEDCOM electrical load reduction measures were implemented or programmed in FY 2007. Several initiatives (such as chilled water storage to shift demand) were considered and discussed with host installations; however, no measures were implemented during 2007.

IV. Data Tables and Inventories

A. FY 2007 Annual Energy Management Data Report – Separate Cover

B. Excluded Facilities Inventory

DoN)

DOE guidance for implementing EPAct05 allows excluding mission critical facilities such as cold iron support to ships, and energy consumed by transmitters and simulators. The guidelines also permit excluding private party consumption that the government has no control of. The following are DoN exclusions:

Exemption	Installation Name
Cold Iron	NSB NEW LONDON CT
Cold Iron	NSY NORFOLK VA
Cold Iron	NAVFAC MID-ATLANTIC, VA
Cold Iron	LANTORDCOM DET CHARLESTON SC
Cold Iron	NAS KEY WEST FL
Cold Iron	NAVSUPACT PORTSMOUTH, NH
Cold Iron	NSB KINGS BAY GA
Cold Iron	NAVSURFWARCEN COASTSYSTA PANAMA CITY FL
Cold Iron	NAB LITTLE CREEK VA
Cold Iron	NAVSTA NEWPORT RI
Cold Iron	NAVSTA ROTA SP
Cold Iron	NAVSTA INGLESIDE TX
Cold Iron	NAVBASE SAN DIEGO CA
Cold Iron	NAVBASE CORONADO SAN DIEGO CA
Cold Iron	NAVSHIPYD PUGET SOUND WA
Cold Iron	NAVMAG INDIAN ISLAND W
Cold Iron	COMFLEACT CHINHAE KS
Cold Iron	COMFLEACT YOKOSUKA JA
Cold Iron	NAVBASE GUAM
Cold Iron	COMFLEACT SASEBO J
Cold Iron	NAVSTA PEARL HARBOR HI
Cold Iron	NAVBASE POINT LOMA, C
Cold Iron	SUBASE BANGOR WA
Cold Iron	NAVSTA EVERETT WA
Cold Iron	NAVBASE VENTURA COUNTY, C
Simulator	LANTORDCOM DET CHARLESTON SC

Exemption	Installation Name
Simulator	NAVBASE SAN DIEGO CA
Simulator	NAVBASE CORONADO SAN DIEGO CA
Simulator	NAS KINGS VILLE TX
Simulator	NAS LEMOORE CA
Simulator	NAVSURFWARCENDIV PORT HUENEME CA
Simulator	NAVBASE POINT LOMA, CA
Simulator	MCAS MIRAMAR, CA
Simulator	NAVAIRENGCEN LAKEHURST NJ

TMA

BUMED excluded the following facilities in its installation facility inventory.

NH Bremerton, WA

B04, Water Storage Tank – Structures such as outside parking garages which consume essentially only lighting energy, yet are classed as buildings. While great opportunity exists for energy efficiency measures in such structures, the relatively lower energy use per square foot of the unconditioned space should not be counted as "building" space. Energy used for airport and street lighting not directly associated with a building may also be assumed to be excluded from goals.

BFH5, Fleet Hospital Training Site (Tents)- Buildings where energy usage is skewed significantly due to reasons such as: buildings entering or leaving the inventory during the year, buildings down- scaled operationally to prepare for decontamination, decommissioning and disposal, and buildings undergoing major renovation and/or major asbestos removal. These buildings may be excluded only during the fiscal year when energy usage is skewed. Energy efficiency measures should be considered for down-scaled buildings awaiting disposal if the measures pay for themselves prior to building demolition.Air Force Listing of Exempt Facilities

Base	Type	Energy	Billion Btu's	KSF	Reason
Arnold	AEDC Cooling water sys	Electricity	101	66	Cooling water supports testing equipment and is not significantly influenced by normal energy conservation methods.
Eglin	Climatic Lab	Electricity	56	203	
Dover	Approach Control	Electricity	0	7	
Dover	Telecom	Electricity	5	24	

(Continued).

Base	Type	Energy	Billion Btu's	KSF	Reason
Dover	Test Cell	Electricity	0	5	This is a national security function. Facility used for equipment testing and is not significantly influenced by normal energy conservation methods. This is a national security function. Radar and communications equipment is not significantly influenced by normal energy conservation methods. This is a national security function. Facility used for communications equipment and is not significantly influenced by normal energy
MacDill	Satellite Communications	Electricity	1	5	
MacDill	FAA Radar	Electricity	1	0	
MacDill	Communications facility	Electricity	4	43	
Travis	Telecom	Electricity	0	7	
611 ASG	Radar and Support	Electricity	47	135	
	Total		215	496	

C. List of Acronyms/Symbols Used in Body of Report

Acronym/Symbol	Meaning
ACC	Air Combat Command
ACCESS	ACC Energy Saving Solutions
AEE	Association of Energy Engineers
AETC	Air Education and Training Command
AFB	Air Force Base
AFCESA	Air Force Civil Engineer Support Agency
AFRPA	Air Force Real Property Agency
AMI	Advanced Metering Infrastructure
ASHRAE	American Society of Heating, Refrigeration and Air Conditioning Engineers
ASN(I&E)	Assistant Secretary of the Navy (Installations and Environment)
BMP	Best Management Practices
BOSC	Base Operations Services Contracts
Btu	British Thermal Unit
BUMED	Bureau of Medicine and Surgery
CCAFS	Cape Canaveral Air Force Station
CDC	Central Distribution Center

CEM	Certified Energy Manager
CEU	Continuing Education Unit
CFL	Compact Fluorescent Lighting
CNIC	Commander Navy Installations Command
DASA(I&H)	Deputy Assistant Secretary of the Army for Installations and Housing
DCMA	Defense Contract Management Agency
DDC	Direct Digital Control
DeCA	Defense Commissary Agency
DFAS	Defense Finance and Accounting Service
DFD	Defense Facilities Directorate
DIA	Defense Intelligence Agency
DIAC	Defense Intelligence Analysis Center
DLA	Defense Logistics Agency
DoA	Department of the Army
DoAF	Department of the Air Force
DoD	Department of Defense
DOE	Department of Energy
DoN	Department of Navy
DUSD(I&E)	The Deputy Under Secretary of Defense (Installations and Environment)
ECIP	Energy Conservation Investment Program
EMCS	Energy Management Control Systems
EMSG	Senior Energy Management Steering Group
EO	Executive Order
EPAct	Energy Policy Act of
ESPC	Energy Savings Performance Contracting
ETSD	Engineering and Technical Services Division
FEMP	Federal Energy Management Program
FES	Facility Energy Supervisor
FOB2	Federal Office Building #2
FY	Fiscal Year
GPC	Government Purchase Card
GovEnergy	Premier Energy Training Workshop and Exposition for Federal Agencies
GSA	General Services Administration
GSF	Gross Square Feet
GSHP	Ground Source Heat Pump
HCFC	Hydro chlorofluorocarbons

HQ	Headquarters
HVAC	Heating, Ventilating and Air Conditioning
IMCOM	Installation Management Command
kW	kilowatt
LCCA	Life-Cycle Cost Analysis
LEED	Leadership in Energy and Environmental Design
LFD	Lease Facilities Division
MAJCOM	Major Command
MEBI	Model Energy Base Initiative
MEDCOM	United States Army Medical Command
MEO	Most Efficient Organization
MILCON	Military Construction
MTF	Military Medical Treatment Facilities
MW	Megawatt
MWh	Megawatt hour
NAVFAC	Naval Facilities Engineering Command
NGA	National Geospatial-Intelligence Agency
NH	Naval Hospital
NMC	Naval Medical Center
NSA	National Security Agency
ORNL	Oak Ridge National Laboratory
OMB	Office of Management and Budget
PACAF	Pacific Air Forces
PBMO	Pentagon Building Management Office
PENREN	Pentagon Renovation Office
PH&RP	Pentagon Heating & Refrigeration Plant
PNNL	Pacific Northwest National Laboratory
PV	Photovoltaics
QSR	Quality Surveillance Representative
RDF	Refuse Derived Fuel
REC	Renewable Energy Certificate
REM	Resource Efficiency Manager
RFP	Request for Proposal
RMCS	Refrigeration Monitoring and Control Systems
SAF/IE	Secretary of the Air Force for Installations, Environment and Logistics
SDD	Sustainable Design and Development
SECAF	Secretary of the Air Force
SECNAV	Secretary of the Navy

SFG	Senior Focus Group
SGS	Space Gateway Support
SIR	Savings-to-Investment Ratio
TMA	Tricare Management Agency
UESC	Utility Energy Services Contracts
USACE	U.S. Army Corp of Engineers
USAF	United States Air Force
USAFA	United States Air Force Academy
USAFE	United States Air Forces in Europe
USGBC	United States Green Building Council
UTF	Utilities Energy Efficiency Task Forces
WHS	Washington Headquarters Service

In: Defense Energy Management
Editor: Edward R. Myers, pp. 93-109
ISBN: 978-1-60692-574-4

Chapter 2

DEFENSE MANAGEMENT: OVERARCHING ORGANIZATIONAL FRAMEWORK COULD IMPROVE DOD'S MANAGEMENT OF ENERGY REDUCTION EFFORTS FOR MILITARY OPERATIONS*

William M. Solis

Mr. Chairman and Members of the Subcommittee:

I am pleased to be here today to discuss the Department of Defense's (DOD) efforts to manage and reduce its demand for mobility energy—that is, the energy required for moving and sustaining its forces and weapons platforms for military operations. Mobility energy accounts for about three-fourths of DOD's total energy consumption.[1] U.S. military forces, for example, require vast quantities of fuel to operate combat and support vehicles; generate power at forward-deployed locations; and move troops, equipment, and supplies. As the single largest energy consumer in the United States, DOD incurs billions of dollars each year in fuel costs, and these costs have been rising in recent years as oil prices have increased. DOD recognizes that its high energy demand presents significant risks to its military forces. Moreover, a February 2008 Defense Science Board report concluded that DOD's high fuel demand compromises operational capability and mission success, requires an excessive logistics infrastructure, creates more risk for support operations than

* Excerpted from GAO Report GAO-08-523T, dated March 13, 2008.

necessary, and increases life cycle operations and support costs.[2] In addition, the report notes that changing a culture that considers energy cheap and abundant is one of the most difficult challenges facing the department and the nation.

Today I would like to provide our perspectives on (1) energy issues that are likely to affect DOD in the future, (2) key departmental and military service efforts to reduce demand for mobility energy, and (3) DOD's management approach to guide and oversee these efforts. This statement is based primarily on the work we conducted for a report that we issued today that addresses DOD's management of energy reduction efforts for military operations.[3] As part of this work, we reviewed several DOD- sponsored studies that have recommended actions DOD could take to better manage its mobility energy challenges.[4] We have also had an opportunity to review the February 2008 Defense Science Board report. We conducted this performance audit from September 2007 through March 2008 in accordance with generally accepted government auditing standards. Those standards require that we plan and perform the audit to obtain sufficient, appropriate evidence to provide a reasonable basis for our findings and conclusions based on our audit objectives. We believe that the evidence obtained provides a reasonable basis for our findings and conclusions based on our audit objectives.

ABSTRACT

Several issues, such as rising fuel costs, worldwide energy demand, and the high fuel burden during operations, underscore the importance of energy to DOD. Fuel costs for DOD are substantial and the volatility of world oil prices will likely continue to affect the department—which may require DOD to make difficult trade-offs, such as redirecting funds from ongoing programs to pay for needed fuel. In addition, both the Army and Marine Corps have plans to grow their forces over the next several years, which will inevitably require larger amounts of fuel to sustain these forces and their weapons systems. Other energy issues that are likely to affect DOD in the future are the increased U.S. dependence on foreign oil, projected increases in the worldwide demand for oil, and uncertainties about world oil supplies. Furthermore, DOD's high fuel requirements on the battlefield can place a significant logistics burden on military forces; limit the range and pace of operations; and add to mission risks, including exposing supply convoys to attack. Given these issues, DOD must be well positioned to effectively manage energy demands for military operations.

DOD and the military services have several initiatives under way to reduce demand for mobility energy. At the department level, the Office of the Secretary of Defense (OSD) created a task force in 2006 to address energy security concerns. Moreover, in 2007, the Deputy Secretary of Defense included energy in DOD's list of the top 25 transformational priorities for the department as part of its initiative to pursue targeted acquisition reforms. Each of the military services also has its own initiatives under way to reduce mobility energy demand. The Army is addressing fuel consumption at forward-deployed locations by developing foam-insulated tents and temporary dome structures that are more efficient to heat and cool, reducing the demand for fuel-powered generators. The Navy has established an energy conservation program to encourage ships to reduce energy consumption. The Air Force has developed an energy strategy and undertaken initiatives to determine fuel-efficient flight routes, reduce the weight on aircraft, optimize air refueling, and improve the efficiency of ground operations. The Marine Corps has initiated research and development efforts to develop alternative power sources, such as hybrid power, and improve fuel management.

While these and other individual efforts are under way to reduce mobility energy demand, we found that DOD does not have an overarching organizational framework to guide and oversee these efforts. Our prior work has shown that an overarching organizational framework is critical to successful transformation in both public and private organizations. Key elements of such a framework include (1) top-level leadership and an implementation team with dedicated resources and funding; (2) a comprehensive strategic plan that includes goals, objectives, methods, timelines, and outcome-oriented performance metrics; and (3) a communication strategy that provides shared expectations and reports related progress. We found that DOD's current approach to mobility energy lacks these elements. For example, while DOD has begun to increase management attention on energy issues, it has not designated a single executive-level OSD official—supported by an implementation team—who is accountable for mobility energy matters across the department, who participates in top policy-making decisions as an advocate for reducing mobility energy demand, and who serves as a stakeholder in interagency discussions about national energy concerns. Currently, DOD's approach to mobility energy is decentralized, with fuel oversight and management responsibilities diffused among several OSD and military service offices as well as working groups. In addition, until DOD fully develops and implements a comprehensive strategic plan for mobility energy, it cannot be certain that mobility energy reduction efforts align with the department's energy mission or strategic goals to ensure that they are appropriately prioritized or to know whether critical gaps or duplication of efforts exist. Finally, without an effective mechanism to facilitate communication of mobility energy reduction efforts among OSD and the military services, DOD cannot be assured that these efforts are consistent with DOD's energy priorities and goals. We also found that DOD has made limited progress in incorporating fuel efficiency as a consideration in key business processes—which include developing requirements for and acquiring new weapons systems—and in implementing recommendations from department-sponsored studies on fuel reduction. With a mobility energy overarching organizational framework in place, DOD would be better positioned to reduce its significant reliance on petroleum-based fuel and to address the energy challenges of the 21st century.

Several Issues Underscore Importance of Energy to DOD

Several issues, such as rising fuel costs, worldwide energy demand, and the high fuel burden during operations, underscore the importance of energy to DOD. Fuel costs for DOD are substantial and the volatility of world oil prices will likely continue to affect the department. For example, in fiscal year 2007, DOD reported that it consumed almost 4.8 billion gallons of mobility fuel and spent $9.5 billion. Although fuel costs represent less than 3 percent of the total DOD budget, they have a significant impact on the department's operating costs. DOD has estimated that for every $10 increase in the price of a barrel of oil, DOD's operating costs increase by approximately $1.3 billion. Furthermore, during a 2007 military energy security forum, DOD officials discussed the possibility of oil prices rising to as much as $200 a barrel if a major disruption were to occur. Rising fuel costs may require DOD to make difficult trade-offs, such as redirecting funds from ongoing programs to pay for needed fuel. In addition, both the Army and Marine Corps have plans to grow their forces over the next several years, which will inevitably require larger amounts of fuel to sustain these forces and their weapons systems.

Other energy issues that are likely to affect DOD in the future are the increased U.S. dependence on foreign oil, projected increases in the worldwide demand for oil, and uncertainties about world oil supplies. In 2007, about 67 percent of the oil consumed in the United States was imported, and the increased energy dependence on other countries raises concern about instability in the Middle East and elsewhere.[5] In addition, the Department of Energy projects that worldwide oil demand will continue to grow, reaching 118 million barrels per day in 2030, up from 84 million barrels per day in 2005. Although countries such as China and India will generate much of this increased demand, the United States will remain the world's largest oil consumer. Moreover, more than 60 percent of world oil reserves are in countries where relatively unstable political conditions could constrain oil exploration and production. Furthermore, worldwide supplies of oil from conventional sources remain uncertain. U.S. oil production peaked around 1970, and worldwide production could peak and begin to decline. Although there is great uncertainty about when this might happen, most studies estimate that oil production will peak sometime between now and 2040.[6] These issues, as well as the increasing threat of climate change, may lead to global instabilities that could require DOD to conduct operations in some of these regions and protect oil supply routes and critical infrastructure—all of which would ultimately lead to increased fuel requirements for the department.

In addition, DOD's high fuel requirements on the battlefield can place a significant logistics burden on military forces, limit the range and pace of operations, and add to mission risks. For example, for current operations, the fuel logistics infrastructure requires, among other things, long truck convoys that move fuel to forward-deployed locations while being exposed to potential enemy attacks. Combatant commanders may also face additional risks related to fuel disruptions in operations. For instance, according to a U.S. Central Command official, changes in customs procedures, truck driver strikes, refinery maintenance, road construction, and holiday periods may close border crossings for long periods of time, possibly resulting in the interruption of fuel supplies to forward-deployed locations. Moreover, a 2007 LMI report stated that the department's increasing fuel demand limits its ability to establish a more mobile and agile force.

DOD and the Military Services Have Made Efforts to Reduce Mobility Energy Demand

DOD and the military services have made efforts to reduce mobility energy demand for their forces and in their weapons systems. At the department level, OSD created the DOD Energy Security Task Force in 2006— consisting of an integrated product team, several working groups, and a senior steering group—to address long-term energy security concerns. Among other activities, the task force is monitoring the progress of selected military service-led research and development projects (see table 1) that have the potential for reducing mobility energy demand.

In addition to focusing on research and development initiatives, DOD has recognized a need to factor energy efficiency considerations into its acquisition process. In 2007, the Deputy Secretary of Defense included energy in DOD's list of the top 25 transformational priorities for the department as part of its initiative to pursue targeted acquisition reforms. Also in 2007, the Office of the Under Secretary of Defense for Acquisition, Technology, and Logistics established a DOD policy to include the fully burdened cost of fuel—that is, the total ownership cost of buying, moving, and protecting fuel in systems during combat—for the acquisition of all tactical systems that create a demand for energy.[7] To incorporate the fully burdened cost of energy into acquisition decisions, OSD initiated a pilot program that includes three systems: the Army and Marine Corps' Joint Light Tactical Vehicle, the Navy's new CG(X) cruiser, and the Air Force's Next-Generation Long-Range Strike aircraft.

Table 1. Selected Energy-Related Research and Development Projects Being Monitored by DOD's Energy Security Task Force

Category	Project name	Description
Air platforms	Highly Efficient Embedded Turbine Engine	Develop a variable core engine to reduce fuel consumption in unmanned aerial vehicles, transport aircraft, and other aircraft.
	Small Heavy Fueled Engine	Extend the duration of unmanned aerial vehicle engines from 3-4 to 6-8 hours to increase fuel efficiency and reduce the logistics tail by using a single battlefield fuel; plan to apply to mobile ground power generators.
	Long-Range Unmanned Aerial Vehicles	Extend flight time of unmanned aerial vehicles for up to 6-7 days for increased fuel efficiency and savings over conventional surveillance and reconnaissance platforms.
Ground vehicles	Fuel-Efficient Ground Vehicle Demonstrator	Identify opportunities in fuel-efficient technologies to build a virtual vehicle that will demonstrate decreased fuel consumption in a tactical vehicle without decreasing performance or capability.
Power systems	Fuel Cell Research	Develop and demonstrate compact and mobile fuel cell systems to provide onboard power generation for increasing power demands and to reduce battery weight.
	Transportable Hybrid Electric Power Supply	Provide hybrid electric power generators to reduce diesel fuel usage and resupply requirements.
	Hybrid Intelligent Power	Automate generators on the battlefield to turn on and off as needed to minimize fuel use and reduce maintenance needs, personnel requirements, and power interruptions.

Source: DOD.

In another initiative, the Joint Staff added language to its guidance in May 2007 requiring that an energy efficiency key performance parameter be selectively considered in the development of capability needs for new systems.[8] The guidance defines a key performance parameter as an attribute or characteristic of a system that is considered critical or essential to the development of an effective military capability.

In addition, each of the military services has its own initiatives under way to reduce mobility energy demand. The following highlights several key efforts and is not intended to be a comprehensive listing of all fuel reduction efforts.

Army

The Army is addressing fuel consumption at forward-deployed locations by developing foam-insulated tents and temporary dome structures that are more efficient to heat and cool and therefore could reduce the demand for fuel-powered generators at these locations. Another initiative is the development of a transportable hybrid electric power station, which uses wind, solar energy, a diesel generator, and storage batteries to provide reliable power with fewer fuel requirements.

Navy

The Navy has established an energy conservation program aimed at encouraging ships to reduce energy consumption. The energy conservation program provides training materials, such as a shipboard energy conservation manual and a pocket guide to assist commanders with energy-saving activities. The program also gives quarterly awards to ships that use less than the Navy's established baseline amount of fuel. The Navy has also made ship design alterations to reduce fuel demand.

Air Force

The Air Force has identified and begun to implement initiatives aimed at reducing mobility energy demand and increasing fuel efficiency, aligning these initiatives with its energy strategy. These initiatives include determining fuel-efficient flight routes, reducing the weight on aircraft, optimizing air refueling, and improving the efficiency of ground operations. In addition, it is testing synthetic fuels in its aircraft that could partly displace the use of petroleum-based fuel.

Marine Corps

The Marine Corps has initiated efforts to develop alternative power sources and improve fuel management. For example, it is testing the use of hybrid power—by combining solar panel, generator, and battery energy sources—at remote sites to lessen its fuel transportation demands to forward-deployed locations. In addition, the Office of Naval Research is leading efforts for the Marine Corps to develop decision support tools that process and analyze data and improve fuel management in combat.

DOD Has Not Established an Overarching Organizational Framework to Guide and Oversee Mobility Energy Reduction Efforts

While DOD and the military services have several efforts under way to reduce mobility energy demand, DOD has not established an overarching organizational framework to guide and oversee these efforts. In the absence of a framework for mobility energy, we also found that DOD has made limited progress in incorporating fuel efficiency considerations into its key business processes and in implementing recommendations from department-sponsored studies on fuel reduction. In the report that we issued today, we made recommendations that DOD establish an overarching organizational framework for mobility energy. Without such a framework, DOD cannot be assured that its current mobility energy reduction efforts will be fully implemented and will significantly reduce its reliance on petroleum-based fuel.

DOD Lacks Key Elements of an Overarching Organizational Framework

Our prior work has shown that an overarching organizational framework is critical to successful transformation in both public and private organizations. The key elements of such a framework include (1) top-level leadership and an implementation team with dedicated resources and funding; (2) a comprehensive strategic plan, including goals and objectives, methods and timelines for evaluating progress, and outcome- oriented performance metrics; and (3) a communication strategy that involves creating shared expectations and reporting related progress.[9] We found that DOD's current approach to mobility energy lacks these elements.

Top-Level Leadership and Implementation Team

While DOD has begun to increase management attention and has identified energy as a transformational priority, it has not designated a single executive-level OSD official whose primary focus is on mobility energy and who is accountable for these matters across the department. Our prior work has stated that leadership must set the direction, pace, and tone and provide a clear, consistent rationale that brings everyone together behind a single mission.[10] Currently, DOD's approach to mobility energy is decentralized, with fuel oversight and management responsibilities diffused among several OSD and military service offices as well as working groups. DOD directives designate the Under Secretary of Defense for Acquisition, Technology, and Logistics as the department's senior energy official, with responsibility for establishing policies, granting waivers, and approving changes in the management of energy commodities, including petroleum.[11] However, it is unclear the extent to which the Under Secretary or any official from this office provides comprehensive guidance and oversight and sets a direction for mobility energy reduction efforts across the department. In addition, the Under Secretary has a broad range of other responsibilities that include, among other things, matters relating to the DOD acquisition system, research and development, systems engineering, logistics, installation management, and business management modernization. Therefore, the Under Secretary's primary focus has not been on the management of mobility energy.

In addition, DOD's Energy Security Task Force was formed to address long-term energy security concerns, such as DOD's reliance on fossil fuels, but we found that the task force has been unable to develop policy or provide guidance and oversight of mobility energy issues across the department. As indicated in its charter, the task force is required to develop a comprehensive DOD energy strategy and an implementation plan. Among other deliverables, the charter also requires it to define DOD's energy challenge, create a compendium of energy-related works, and perform a strategic assessment of energy. While the task force has taken steps to identify and monitor the progress of selected mobility energy reduction projects across the department, it has not yet completed an energy strategy or implementation plan, as well as other responsibilities. Furthermore, OSD officials told us that while the task force has briefed the Deputy Secretary of Defense's advisory group on its recommended projects, it does not have a "seat at the table" in departmental discussions at the Deputy Secretary of Defense level or at other executive levels, such as the Joint Requirements Oversight Council, the Defense Acquisition Boards, or the 3-Star Group within DOD's Planning, Programming, Budgeting, and Execution process, as an advocate for reducing mobility energy demand.[12]

DOD also does not have an implementation team in place, with dedicated resources and funding, to address mobility energy issues. For example, the officials who lead DOD's

Energy Security Task Force's integrated product team do so as an extra responsibility outside of their normal work duties. Other DOD officials said that the task force provides a good forum for sharing energy ideas across the department but lacks adequate staff to carry out specific actions. Furthermore, a task force participant told us that it can be difficult to find time to attend meetings while balancing other duties. The task force also does not receive any dedicated funding to pursue department-level energy priorities. Without a long-term funding mechanism, DOD may not be able to ensure that mobility energy reduction efforts receive sustained funding over a period of years.

Moreover, DOD may not be well positioned to serve as a focal point on mobility energy within the department, with Congress, and with the Department of Energy or other interagency partners. During a military energy security forum held at the National Defense University in November 2007, representatives from various DOD offices presented energy as an area that is significant to a breadth of issues ranging from force protection to global stability to the security of DOD's critical infrastructure. They also noted that DOD has the potential to play multiple roles with respect to energy, including consumer, market leader, educator/motivator, oil infrastructure protector, and warfighter supporter. These concerns, coupled with an increased national and congressional interest in reducing fossil fuel dependence and exploring alternative energies, will likely necessitate an increased leadership focus on long-term energy issues, both within DOD and in its role as a stakeholder in interagency and national dialogues.

Comprehensive Strategic Plan

DOD has not yet developed a comprehensive strategic plan for mobility energy, although it has taken some steps to lay the foundation for mobility energy strategic planning. According to OSD officials, DOD has begun to incorporate mobility energy issues into its *Guidance on the Development of the Force,* a department-level strategic planning document. In addition, the Office of the Deputy Assistant Secretary of Defense for Policy Planning, within the Office of the Under Secretary of Defense for Policy, is analyzing future energy concerns for the United States and the international security environment and highlighting their implications for the department. DOD officials said that the analysis is expected to provide information for consideration in the development of future strategic planning documents. We also observed that the DOD Energy Security Task Force has begun efforts to define goals that eventually may be incorporated into a DOD energy security strategic plan. OSD officials told us that the task force's intent is to complete this strategic plan by May 2008. However, current DOD strategic planning documents, such as the *National Military Strategy* and the most recent *Quadrennial Defense Review*, do not address mobility energy reduction.[13] Furthermore, until DOD fully develops and implements a comprehensive strategic plan for mobility energy, it cannot be certain that mobility energy reduction efforts align with the department's energy mission or strategic goals, ensure that they are appropriately prioritized, or know whether critical gaps or duplication of efforts exist.

Communication Strategy

DOD does not have an effective mechanism to facilitate communication and coordination of mobility energy reduction efforts among OSD and the military services. In addition, we found a lack of cross-service coordination concerning mobility energy reduction initiatives. While DOD's Energy Security Task Force aims to identify key players within the energy field, its current structure does not ensure departmentwide communication of fuel reduction efforts, particularly among the military services, which are responsible for most of these efforts. More specifically, during our observation of a task force monthly meeting, we found that although this venue provides for some sharing of information, the generally less than 2 hours allotted for each monthly meeting does not allow for effective coverage of the spectrum of DOD's mobility energy issues. Moreover, we noted in our chapter that although the task force's senior steering group includes, among others, the service under secretaries and assistant secretaries; the Director, Defense Research and Engineering; and several principal deputy under secretaries of defense, it only meets two to three times a year. Additionally, with the exception of the Air Force, none of the other military service members on the senior steering group have primary responsibility for mobility energy efforts within their services. Without executive-level focal points, the military services may not be well positioned to effectively coordinate on mobility energy efforts across the department or provide leadership or accountability for efforts within their services. Furthermore, DOD cannot be assured that energy reduction efforts are consistent with DOD's energy priorities and goals.

DOD Has Made Limited Progress in Incorporating Fuel Efficiency into Key Business Processes and in Implementing Recommendations from Department-Sponsored Studies

In the absence of an overarching organizational framework, DOD has made limited progress in incorporating fuel efficiency as a consideration in key business processes—which include developing requirements for and acquiring new weapons systems—and in implementing recommendations made in department-sponsored studies on fuel reduction.

DOD Has Not Yet Fully Incorporated Fuel Efficiency Considerations into Its Key Business Processes

While DOD has recently begun to take some steps to integrate fuel considerations into its key departmental business processes, such as its requirements development and acquisition processes for new weapons platforms and other mobile defense systems, these considerations are not factored in a systematic manner and cannot be fully applied. For example, DOD's requirements development process does not systematically include energy efficiency considerations, and the capability gap assessments associated with the process do not include fuel-related logistics, thus leaving these types of issues to be resolved after systems are fielded. In May 2007, the Joint Staff established an energy efficiency key performance parameter that would require fuel considerations during capabilities development. However, because DOD has not developed a methodology to determine how best to employ the energy efficiency key performance parameter, its implementation remains uncertain.

DOD has also taken steps to inform its acquisition process with its pilot program to determine the fully burdened cost of fuel for three mobile defense systems. While the pilot program represents a step toward providing visibility over the total logistics costs associated with delivered fuel and DOD has set a fall 2008 deadline to issue guidance for applying the fully burdened cost of fuel in acquisition programs, DOD has not yet developed an approach for determining how it would incorporate this information into its acquisition decision-making process. Until the pilot program is completed and the results are assessed, DOD is not in a position to apply a fully burdened cost analysis to its acquisition process. Thus, the department is unable to promote greater visibility over its acquisition decisions or more fully consider the operational and cost consequences of the fuel burden on the logistics infrastructure.

Other key DOD business processes, such as those that address repair, recapitalization, and replacement of mobile defense systems, also present opportunities to incorporate fuel efficiency measures during system upgrades. However, OSD officials told us that the department generally makes decisions about system upgrades without regard to fuel efficiency, including the fully burdened cost of fuel, in part because such decisions require greater up-front costs. Although DOD recognizes that by reducing energy demand it can provide its forces greater flexibility and reduce their dependency on the logistics infrastructure, some OSD officials told us that DOD's budget process promotes a short-term outlook and does not encourage the purchase of fuel-efficient systems or upgrades that may initially cost more but could reduce life cycle and logistics costs over the long term. Moreover, the 2008 Defense Science Board report noted that DOD's lack of tools to assess the operational and economic benefits of fuel efficiency technologies is a major reason why DOD underinvests in the development and deployment of these technologies.

DOD Has Been Slow to Implement Recommendations from Department-Sponsored Studies on Fuel Reduction

In the absence of an overarching organizational framework, DOD has made limited progress in implementing recommendations from department-sponsored studies by organizations such as the Defense Science Board, The JASONs, and LMI that have urged an expansion of efforts to reduce dependency on petroleum-based fuel. These studies confirmed that for many reasons, continued heavy reliance on petroleum- based fuel poses a significant problem for DOD. For example, LMI reported that DOD's increasing fuel demand furthers the nation's reliance on foreign energy sources and, as we mentioned previously, limits the department's ability to establish a more mobile and agile force. The studies found a need to focus more DOD management attention on mobility energy matters and recommended actions aimed at, among other things, improving the fuel efficiency of weapons platforms, eliminating institutional barriers that bear upon the department's decisions regarding fuel efficiency, and developing a long-term mobility energy strategy that would lead to reduced consumption of petroleum-based fuel.

DOD has not taken a formal position on these recommendations, and implementation, in some cases, would require significant changes throughout the department that could generate institutional resistance. One study, for example, called for creating a unified energy governance structure in order to alter DOD's "energy culture." During our review, we found that DOD had taken some steps toward implementing some of the recommendations, such as initiating a pilot program for determining the fully burdened cost of delivered fuel and adding a requirement for a fuel efficiency key performance parameter in its Joint Staff policy manual. However, other recommendations, such as establishing a governance structure for mobility energy, have not been implemented. Furthermore, the 2008 Defense Science Board report noted that the recommendations made by the 2001 Defense Science Board report are still open and remain viable. Our chapter, which was issued today, presented the recommendations from these department-sponsored studies and actions DOD has taken in more detail. We also concluded that an overarching organizational framework could better position DOD to address these and other fuel reduction recommendations in a more timely and effective manner.

Overarching Organizational Framework Needed to Better Position DOD to Address Mobility Energy Challenges

In the chapter we issued today, we recommended that DOD establish an overarching organizational framework for mobility energy to improve the department's ability to guide and oversee mobility energy reduction efforts.[14] To establish such a framework, DOD should designate an executive-level OSD official who is accountable for mobility energy matters; develop a comprehensive, departmentwide strategic plan; and improve DOD's business processes to incorporate energy efficiency considerations. In addition, we recommended that the military services designate executive-level focal points to establish effective

communication and coordination among OSD and the military services on departmentwide mobility energy reduction efforts as well as to provide leadership and accountability over their own efforts. With a mobility energy overarching organizational framework in place, DOD would be better positioned to reduce its significant reliance on petroleum-based fuel and to address the energy challenges of the 21st century. In commenting on a draft of our chapter, DOD partially concurred with our recommendations.

Mr. Chairman and Members of the Subcommittee, this concludes my prepared statement. I would be happy to answer any questions that you may have at this time.

References

[1] Energy consumed at fixed installations, referred to as facility energy, accounts for most of DOD's remaining energy use.

[2] Defense Science Board Task Force on DOD Energy Strategy, *More Fight—Less Fuel* (February 2008).

[3] GAO, Defense Management: Overarching Organizational Framework Needed to Guide and Oversee Energy Reduction Efforts for Military Operations, GAO-08-426 (Washington, D.C.: Mar. 13, 2008).

[4] Defense Science Board Task Force on Improving Fuel Efficiency of Weapons Platforms, *More Capable Warfighting Through Reduced Fuel Burden* (January 2001). The JASONs, *Reducing DOD Fossil-Fuel Dependence*, JSR-06-135 (September 2006). LMI, *Transforming the Way DOD Looks at Energy: An Approach to Establishing an Energy Strategy*, Report FT602T1 (April 2007).

[5] GAO, Department of Energy: Oil and Natural Gas Research and Development Activities, GAO-08-190R (Washington, D.C.: Nov. 6, 2007).

[6] For a discussion of issues surrounding peak oil production, see GAO, Crude Oil: Uncertainty about Future Oil Supply Makes It Important to Develop a Strategy for Addressing a Peak and Decline in Oil Production, GAO-07-283 (Washington, D.C.: Feb. 28, 2007).

[7] Office of the Under Secretary of Defense for Acquisition, Technology, and Logistics Memorandum, "Fully Burdened Cost of Fuel Pilot Program," April 10, 2007.

[8] Joint Chiefs of Staff Instruction 3170.01F, Joint Capabilities Integration and Development System (May 1, 2007) and Joint Chiefs of Staff Manual 3170.01C, Operation of the Joint Capabilities Integration and Development System (May 1, 2007).

[9] GAO, Results-Oriented Cultures: Implementation Steps to Assist Mergers and Organizational Transformations, GAO-03-669 (Washington, D.C.: July 2, 2003), and Agencies' Strategic Plans Under GPRA: Key Questions to Facilitate Congressional Review, GAO/GGD-10.1.16 (Washington, D.C.: May 1997).

[10] GAO-03-669.

[11] DOD Directive 4140.25, DOD Management Policy for Energy Commodities and Related Services (Apr. 12, 2004), and DOD Directive 5134.01, Under Secretary of Defense for Acquisition, Technology, and Logistics (Dec. 9, 2005).

[12] The 3-Star Group within DOD's Planning, Programming, Budgeting, and Execution process includes members from OSD's Director of Program Analysis and Evaluation; OSD's under secretaries of defense; the Joint Staff Director for Structure, Resources, and Assessment; and the military services' 3-Star programmers. This group addresses major issues and presents decision options to the Secretary of Defense.

[13] The *National Military Strategy*, signed by the Chairman of the Joint Chiefs of Staff, is guided by the goals and objectives contained in the present *National Security Strategy* and serves to implement the Secretary of Defense's *National Defense Strategy*. The *Quadrennial Defense Review*, prepared by the Secretary of Defense every 4 years, assesses the nature and magnitude of the political, strategic, and military risks associated with executing the missions called for under the *National Defense Strategy*.

[14] GAO-08-426.

In: Defense Energy Management
Editor: Edward R. Myers, pp. 111-154
ISBN: 978-1-60692-574-4

Chapter 3

DEFENSE MANAGEMENT: OVERARCHING ORGANIZATIONAL FRAMEWORK NEEDED TO GUIDE AND OVERSEE ENERGY REDUCTION EFFORTS FOR MILITARY OPERATIONS*

U.S. Government Accountability Office

March 13, 2008
The Honorable
Solomon P. Ortiz
Chairman

The Honorable
J. Randy Forbes
Ranking Member
Subcommittee on Readiness Committee on Armed Services
House of Representatives

The Department of Defense (DOD) is the nation's single largest consumer of energy and relies heavily on petroleum-based fuel for mobility energy— that is, the energy required for moving and sustaining its forces and weapons platforms for military operations. U.S. military forces, for example, require vast quantities of fuel to operate combat and support vehicles; generate power at forward-deployed locations; and

* Excerpted from GAO Report GAO-08-426, dated March 2008.

move troops, equipment, and supplies. In 2007, more than 55 million gallons of fuel, on average, were supplied by DOD each month to support the U.S. forces in Iraq and Afghanistan. Mobility energy accounts for about three-fourths of DOD's total energy consumption.[1] DOD incurs billions of dollars each year in fuel costs, and these costs have been rising in recent years as oil prices have increased. Moreover, high fuel requirements on the battlefield can place a significant logistics burden on military forces; limit the range and pace of operations; and add to mission risks, including exposing convoys to attack.

This chapter responds to a request by the Subcommittee on Readiness, House Committee on Armed Services, that we assess DOD's efforts to reduce its reliance on petroleum-based fuel. Specifically, the objectives of this review were to (1) identify key departmental and military service efforts that have been undertaken to reduce demand for mobility energy and (2) assess the extent to which DOD has established an overarching organizational framework to guide and oversee these efforts.

We conducted work at the Office of the Secretary of Defense (OSD); the Joint Staff; the headquarters of the Army, Air Force, Navy, and Marine Corps; and the Defense Logistics Agency. To identify key departmental and military service efforts to reduce mobility energy demand, we reviewed documentation on the objectives and status of ongoing initiatives. In assessing the extent to which DOD has established an overarching organizational framework for mobility energy, we analyzed DOD and military service policies and other documents and reviewed relevant DOD-sponsored studies. We also discussed mobility energy issues with agency officials to gain their perspectives. We conducted our review from September 2007 through March 2008 in accordance with generally accepted government auditing standards. Details on our scope and methodology are contained in appendix I.

RESULTS IN BRIEF

OSD, the Joint Staff, and the military services have undertaken a number of efforts to reduce mobility energy demand. For example:

- OSD established a departmentwide Energy Security Task Force that, among other things, is monitoring the progress of selected energy-related research and development projects. OSD also has begun a pilot program for assessing the full energy costs of new weapons systems rather than just the cost of the fuel itself as part of the acquisition process.[2] Moreover, in 2007, the Deputy Secretary of Defense included energy in DOD's list of the top 25 transformational priorities for the department, as part of its initiative to pursue targeted acquisition reforms.

- The Joint Staff updated its policy governing the development of capability requirements for new weapons systems to require that energy efficiency be selectively considered as a key performance parameter.[3]
- The Army is addressing fuel consumption at forward-deployed locations by developing foam-insulated tents and temporary dome structures that are more efficient to heat and cool and therefore could reduce the demand for fuel-powered generators at these locations.
- The Navy has established an energy conservation program aimed at encouraging ships to reduce energy consumption. It has also made ship design alterations to reduce fuel demand.
- The Air Force has developed an energy strategy and undertaken initiatives to determine fuel-efficient flight routes, reduce the weight on aircraft, optimize air refueling, and improve the efficiency of ground operations. In addition, it is testing synthetic fuels in its aircraft that could partly displace the use of petroleum-based fuel.
- The Marine Corps has initiated efforts to develop alternative power sources and improve fuel management. For example, it is testing the use of hybrid power—by combining solar panel, generator, and battery energy sources—at remote sites to lessen its fuel transportation demands to forward-deployed locations.

While these and other individual efforts are under way to reduce mobility energy demand and DOD has identified energy as one of its transformational priorities, DOD lacks key elements of an overarching organizational framework to guide and oversee these efforts. Our prior work has shown that such a framework is critical to successful transformation in both public and private organizations. In the absence of an overarching organizational framework for mobility energy, DOD cannot be assured that its current efforts will be fully implemented and will significantly reduce its reliance on petroleum-based fuel. More specifically, we found that DOD's current approach to mobility energy lacks (1) top leadership, with a single executive-level OSD official— supported by an implementation team—who is accountable for mobility energy matters; (2) a comprehensive strategic plan for mobility energy that aligns individual efforts with DOD-wide goals and priorities, establishes approaches or strategies to achieve goals, and evaluates progress through performance metrics; and (3) an effective mechanism to provide for communication and coordination of mobility energy efforts among OSD and the military services as well as leadership and accountability over each military service's efforts. We also found that DOD has made limited progress in incorporating fuel efficiency as a consideration in key business processes—which include developing requirements for and acquiring new weapons systems. According to OSD and military service officials, DOD has not established an overarching organizational framework for mobility energy in part because of concerns regarding how such a framework would be implemented, how it would integrate with other existing organizational responsibilities, and how it would affect ongoing efforts to reduce mobility energy demand. However, DOD has created a management framework to oversee facility energy, which accounts for about 25 percent of the department's energy use, and has established new organizational frameworks to address other crosscutting issues, such as business systems modernization, corrosion control and prevention, contractors on the battlefield, and the defeat of improvised explosive devices. The establishment of such a framework for mobility energy could provide greater assurance that DOD's efforts to reduce its reliance on petroleum-based fuel will succeed without degrading its operational capabilities and that DOD is better positioned to address future mobility energy challenges—both within the department and as a stakeholder in national energy security dialogues.

We are recommending that DOD establish an overarching organizational framework for mobility energy to improve the department's ability to guide and oversee mobility energy reduction efforts. To establish such a framework, DOD should designate an executive-level OSD official—with an implementation team—who is accountable for mobility energy matters; develop a comprehensive, departmentwide strategic plan; and improve DOD's business processes to fully incorporate energy efficiency considerations. In addition, we are recommending that the military services designate executive-level focal points to establish effective communication and coordination among OSD and the military services on departmentwide mobility energy efforts as well as provide leadership and accountability over their own efforts. In commenting on a draft of this chapter, DOD partially concurred with our recommendations. DOD's comments are reprinted in appendix III.

BACKGROUND

Energy, and specifically petroleum-based fuel, will be a key issue facing the nation during the 21st century. The United States accounts for only 5 percent of the world's population but about 25 percent of the world's oil demand. The Department of Energy projects that worldwide oil demand will continue to grow, reaching 118 million barrels per day in 2030, up from 84 million barrels per day in 2005. Although countries such as China and India will generate much of this increased demand, the United States will remain the world's largest oil consumer. World oil production has been running at near capacity in recent years to meet rising consumption, putting upward pressure on oil prices. The potential for disruptions in key oil-producing regions of the world, such as the Middle East, and the yearly threat of hurricanes in the Gulf of Mexico have also exerted upward pressure on oil prices. Crude oil prices almost tripled from 2003 through the beginning of 2008, rising from $36 a barrel to as high as $100 a barrel.

In 2007, about 67 percent of the oil consumed in the United States was imported, and the increased energy dependence on other countries raises concern about international turmoil in the Middle East and elsewhere.[4] In addition, worldwide supplies of oil from conventional sources remain uncertain. U.S. oil production peaked around 1970, and worldwide production could peak and begin to decline, although there is great uncertainty about when this might happen.[5] Moreover, there are differences of opinion as to how long the nation can rely on petroleum- based fuel to meet the majority of its energy needs. As a result, we have previously reported that, in addition to expanding production, the United States may need to place more emphasis on demand reduction strategies as well as developing alternative or renewable energy supplies and technologies.[6]

DOD is the single largest energy consumer in the United States, and it consumes about 90 percent of the petroleum-based fuel used by the U.S. government. Jet fuel constitutes more than half of DOD's total energy consumption. Other types of petroleum-based fuels used by DOD include marine and auto diesel. According to the *Department of Defense Annual Energy Management Report* for fiscal year 2006, DOD consumed approximately 4.6 billion gallons of mobility fuels in fiscal year 2006, down from 5.17 billion gallons in fiscal year 2005. However, spending on mobility fuels increased 26.5 percent, from $7.95 billion in fiscal year 2005 to $10.06 billion in fiscal year 2006. DOD attributed this cost increase to the rise in fuel prices. For example, the price of jet fuel increased from $1.70 per gallon in fiscal year 2005 to $2.34 per gallon in fiscal year 2006. Congress, in fiscal year 2006, provided DOD more than $2 billion in supplemental funds to cover increased fuel costs. In fiscal year 2007, DOD reported that the department consumed almost 4.8 billion gallons of mobility fuel and spent $9.5 billion. Although fuel costs represent less than 3 percent of the total DOD budget, they have a significant impact on the department's operating costs. DOD has estimated that for every $10 increase in the price of a barrel of oil, DOD's operating costs increase by approximately $1.3 billion.

Fuel presents an enormous logistical burden for DOD when planning and conducting military combat operations. For current operations, the fuel logistics infrastructure requires, among other things, long truck convoys that move fuel to forward-deployed locations while exposed to the vulnerabilities of operations, such as enemy attacks (see figs. 1 and 2). Army officials have estimated that about 70 percent of the tonnage required to position its forces for battle consists of fuel and water. An armored division can use 600,000 gallons of fuel a day, and an air assault division can use 300,000 gallons a day. In addition, combat support units consume more than half of the fuel the Army uses on the battlefield. Aircraft also burn through fuel at rapid rates; a B-52H, for example, burns approximately 3,500 gallons per flight hour. Of the four military services, the Air Force consumes the greatest amount of petroleum-based fuels.

Source: Headquarters, Marine Corps.

Figure 1. Three-Mile Backup of Fuel Delivery Trucks and Other Supply Vehicles Inside Afghanistan along the Northern Passage from Pakistan (February 2007).

Source: DOD.

Figure 2. A Line of Tanker Trucks Loading Fuel in Kuwait.

DOD has existing policies and organizational responsibilities for managing energy commodities, including petroleum, natural gas, coal, and electricity, to support peacetime and wartime missions and to permit successful and efficient deployment and employment of forces. Its overarching policy directive on managing energy commodities and related services establishes policy on standardizing fuels, minimizing inventory levels, maximizing use of alternative fuel sources from host nations and commercial sources, and privatizing energy infrastructure at military installations.[7] The Defense Energy Support Center, within the Defense Logistics Agency, finances fuel purchases through a defense working capital fund. The military services purchase fuel from the Defense Energy Support Center using funds appropriated for their operation and maintenance accounts. Various DOD components have a role in planning for fuel demand and managing fuel storage and delivery.

DOD has been exploring issues surrounding its reliance on petroleum through a number of studies sponsored by various offices within OSD. In 200 1, the Defense Science Board issued the results of its study on improving the fuel efficiency of weapons platforms, in response to a tasking from the Under Secretary of Defense for Acquisition, Technology, and Logistics.[8] In 2006, the Office of the Director, Defense Research and Engineering, sponsored a study by The JASONs, an independent defense advisory group under The MITRE Corporation, to assess ways to reduce DOD's dependence on fossil fuels.[9] Under the sponsorship of the Office of Force Transformation and Resources, within the Office of the Under Secretary of Defense for Policy, LMI issued a 2007 report on an approach to establishing a DOD energy strategy.[10] During the period in which we were conducting our review, the Defense Science Board, at the direction of the Under Secretary of Defense for Acquisition, Technology, and Logistics, issued a new report on DOD's energy strategy.[11] These studies have been supplemented by internal DOD reviews and other efforts, such as informational forums at the National Defense University, to explore fuel reduction strategies.

Departmental and Military Service Efforts Are under Way to Reduce Mobility Energy Demand

OSD, the Joint Staff, and the military services have made efforts to reduce mobility energy demand for DOD's forces and in weapons platforms. At the department level, OSD and the Joint Staff have several efforts under way to begin to incorporate fuel efficiency considerations in DOD's requirements development and acquisition processes. In addition, each of the military services has its own initiatives under way to reduce mobility energy demand. The discussion that follows highlights several key efforts and is not intended to be a comprehensive listing of all fuel reduction efforts.

OSD and the Joint Staff Have Begun to Address Mobility Energy Demand

Department officials from several offices within OSD and the Joint Staff have initiated efforts to address mobility energy demand. In 2006, OSD created the DOD Energy Security Task Force to address energy security concerns. The task force's integrated product team, which includes representatives from the military services; defense agencies; the Office of the Under Secretary of Defense for Acquisition, Technology, and Logistics; the Office of the Under Secretary of Defense for Policy; the Office of the Principal Deputy Under Secretary of Defense (Comptroller); the Joint Staff; and OSD's Program Analysis and Evaluation office, typically meets each month and has formed several working groups to share information and ideas on efforts to reduce fuel demand in current and future weapons platforms. The integrated product team reports to a senior steering group, consisting of principal deputy secretaries of defense and service under secretaries and assistant secretaries. Among other activities, the task force recommended funding in fiscal year 2008 for several military service-led energy-related research and development projects, and it is monitoring their progress (see table 1).

In addition to focusing on research and development initiatives, DOD has recognized a need to factor energy efficiency considerations into its acquisition process. In 2007, the Deputy Secretary of Defense included energy in DOD's list of the top 25 transformational priorities for the department, as part of its initiative to pursue targeted acquisition reforms. Also, in April 2007, the Office of the Under Secretary of Defense for Acquisition, Technology, and Logistics established a DOD policy to include the fully burdened cost of fuel—that is, the total ownership cost of buying, moving, and protecting fuel in systems during combat—for the acquisition of all tactical systems that create a demand for energy.[12] To incorporate the fully burdened cost of energy into acquisition decisions, OSD initiated a pilot program that includes three systems: the Army and Marine Corps' Joint Light Tactical Vehicle, the Navy's new CG(X) cruiser, and the Air Force's Next-Generation Long-Range Strike aircraft. To further facilitate the implementation of this policy, OSD's Program Analysis and Evaluation office developed a methodology for assessing the fully burdened cost of fuel and completed its initial analyses of the first system, the Joint Light Tactical Vehicle, last fall. According to the DOD policy, the results of the pilot program are expected to be used as the basis for implementation across all relevant acquisition programs.

Table 1. Selected Energy-Related Research and Development Projects Being Monitored by DOD's Energy Security Task Force

Category	Project name	Description
Air platforms	Highly Efficient Embedded Turbine Engine	Develop a variable core engine to reduce fuel consumption in unmanned aerial vehicles, transport aircraft, and other aircraft.
	Small Heavy Fueled Engine	Extend the duration of unmanned aerial vehicle engines from 3-4 to 6-8 hours to increase fuel efficiency and reduce the logistics tail by using a single battlefield fuel; plan to apply to mobile ground power generators.
	Long-Range Unmanned Aerial Vehicles	Extend flight time of unmanned aerial vehicles for up to 6-7 days for increased fuel efficiency and savings over conventional surveillance and reconnaissance platforms.
Ground vehicles	Fuel-Efficient Ground Vehicle Demonstrator	Identify opportunities in fuel-efficient technologies to build a virtual vehicle that will demonstrate decreased fuel consumption in a tactical vehicle without decreasing performance or capability.
Power systems	Fuel Cell Research	Develop and demonstrate compact and mobile fuel cell systems to provide onboard power generation for increasing power demands and to reduce battery weight.
	Transportable Hybrid Electric Power Supply	Provide hybrid electric power generators to reduce diesel fuel usage and resupply requirements.
	Hybrid Intelligent Power	Automate generators on the battlefield to turn on and off as needed to minimize fuel use and reduce maintenance needs, personnel requirements, and power interruptions.

Source: DOD.

In another initiative, the Joint Staff added language to its guidance in May 2007 requiring that an energy efficiency key performance parameter be selectively considered in the development of capability requirements for new systems.[13] The guidance defines a key performance parameter as an attribute or characteristic of a system that is considered critical or essential to the development of an effective military capability. For example, a survivability key performance parameter is applicable for manned systems designed to enhance personnel survival when employed in an asymmetric threat environment. In general, a key performance parameter represents a system attribute that is so significant that failure to meet its minimum threshold could be a reason for DOD or the military services to reevaluate the concept or system or terminate the program.

In response to the work conducted by the DOD Energy Security Task Force, the Joint Staff has also been directed to lead an assessment of simulator capability and capacity across the department. This effort is expected to analyze whether the increased use of simulators could substitute for live training without degrading operational capability. The study will also identify barriers to implementation and needed policy changes.

Army Is Examining Ways to Reduce in-Theater Fuel Demand

The Army has begun a number of efforts to reduce mobility energy demand. These activities include undertaking initiatives to reduce fuel consumption in theater, determining the total costs of delivering fuel, and developing an Army energy strategy. The Army, through the office of the Army Rapid Equipping Force, created the Power Surety Task Force in 2006 to address a joint urgent operational needs statement from a U.S. commander in Iraq that called for alternative energy sources to reduce the amount of fuel transported to supply power generation systems at forward-deployed locations. The Power Surety Task Force aims to foster the development of projects and programs that are deployable within 18 months. Two of the Power Surety Task Force's initiatives—foam-insulated tents and temporary biodegradable dome structures that are more efficient to heat and cool—are expected to reduce the number of generators required to produce power at forward-deployed locations. Another initiative is the development of a transportable hybrid electric power station, which uses wind, solar energy, a diesel generator, and storage batteries to provide reliable power with fewer fuel requirements. According to Army Rapid Equipping Force officials, the power station could potentially replace about half of the current generators at forward- deployed locations. Moreover, they estimated that annual savings in Iraq from some of these initiatives could be at least $1.7 billion, and that other benefits could include a reduction in the number of trucks required in supply convoys, potentially saving lives and reducing vehicle maintenance requirements. We did not validate the Army Rapid Equipping Force's cost savings estimate.

Another ongoing Army activity is its effort to determine the total costs of delivered energy for Army systems. The Army's "Sustain the Mission Project" was started in 2004 to institutionalize a fully burdened cost methodology in the Army. The methodology uses existing Army and DOD databases, metrics, and processes to calculate the fully burdened cost of fuel and to facilitate "what if" analyses for different assumptions and scenarios. It is also aimed at enabling decision makers to perform cost- benefit analyses of investments in alternative energy and weapons systems technologies. The Army has scheduled a demonstration of this tool in late March 2008.

The Army will also sponsor a study that officials expect will lead to the development of a tactical fuel and energy strategy for the future modular force. The contract for the 1-year study was expected to be awarded in 2008. Army officials told us that they plan to update the Army's energy regulation following completion of the study. The current regulation focuses on facility energy, but according to Army officials, the updated version is expected to include mobility fuel as well.

Navy Has Established an Energy Conservation Program and Other Mobility Energy Reduction Initiatives

The Navy has established a shipboard energy conservation program and has undertaken other initiatives to save fuel on ships. The energy conservation program has both training and award components to encourage ships to reduce energy consumption. Training materials and activities include a shipboard energy conservation manual, a pocket guide to assist commanders with energy-saving activities, energy audits of ships to show commanders how energy can be saved, and energy conservation seminars and workshops. Awards are given quarterly to ships that use less than the Navy's established baseline amount of fuel, and fuel savings achieved during the quarter are reallocated to the ship for the purchase of items such as paint, coveralls, and firefighting gear. The ship energy conservation program receives $4 million in funding annually, and Navy officials told us that they achieved $124.6 million in cost avoidance in fiscal year 2006. They said that some other benefits of this program include more available steaming hours, additional training for ships, improved ship performance, reduced ship maintenance, and conservation of resources.

The Navy has undertaken other mobility energy reduction efforts as part of its ship energy conservation program, such as ship alterations. Two key ship alterations are the use of stern flaps and the modification of boiler boxes. A stern flap alters the water flow at the stern to reduce a ship's resistance and increase fuel efficiency. According to Navy officials, preliminary tests of stern flaps on guided missile destroyers showed an annual fuel reduction of 3,800 to 4,700 barrels, or about 6 to 7.5 percent per ship, which DOD estimated would result in potential savings of almost $195,000 per year per ship. Boiler box modifications for amphibious assault ships, one of the Navy's largest fuel-consuming ships, are expected to decrease the amount of fuel expended by 2 percent per ship. Navy officials told us that this alteration has been approved and that most alterations would be completed in fiscal year 2009. According to Navy officials, once all alterations are completed in fiscal year 2011, this effort could potentially save approximately $30 million per year, depending on the price of fuel. We did not validate these potential savings.

Air Force Has Implemented an Energy Strategy and Begun Mobility Energy Reduction Initiatives

In 2005, the Air Force implemented an energy strategy that consists of three components: reducing demand, increasing supply, and changing the culture. At the time of our chapter, the Air Force was in the process of updating its instructions and directives to reflect its energy strategy and to establish an overarching Air Force energy policy. In addition, the Air Force has identified and begun to implement initiatives aimed at reducing mobility energy demand and increasing fuel efficiency, aligning these initiatives with its energy strategy. Four key initiatives are as follows:

- **Direct routing**. This initiative intends to reduce flight time and fuel consumption by flying the most fuel-efficient flight routes and altitudes.
- **Weight reduction**. This initiative intends to decrease excess weight on an aircraft without adversely affecting mission capability. Three categories that are being considered are taking unused items off the aircraft, taking fewer of the items that are needed, and looking at mission-critical items that could be designed differently, for example, with lighter materials. According to Air Force officials, every 100 pounds of weight equate to 1.6 million pounds of fuel, or $686,000 per year across its fleet of mobility aircraft.
- **Air refueling optimization**. With this initiative, the Air Force intends to change the flight planning process to limit air refueling to only when it is mission essential.
- **Efficient ground operations**. This initiative intends to reduce fuel burn during ground operations. Some actions include reducing warm-up time and taxiing on fewer engines.

In addition to these demand-reduction initiatives, the Air Force is pursuing efforts to increase supply through the research and testing of new technologies, as well as renewable and sustainable resources. Through the Air Force's synthetic fuel initiative, jet fuels made from alternative energy sources, such as coal, natural gas, and biomass, are being evaluated for use in military aircraft with the goal of reducing future fuel costs and ensuring fuel availability. The Air Force completed initial testing of a synthetic blend of fuel in the B-52H bomber and certified the use of this fuel blend for this aircraft in August 2007. The service has begun testing on the C-17 cargo aircraft, the B-1 bomber, and the F-22 fighter, with certification expected in 2008. Air Force officials said that they expect the entire fleet to be certified to fly on the synthetic blend of fuel by 2011.

However, our prior work has highlighted challenges associated with the development and adoption of alternative energy sources.[14]

Finally, the Air Force aims to create a culture that emphasizes energy considerations in all of its operations. Air Force officials told us that this component of their strategy has multiple elements, including focused leadership, training, educational curricula, and communication.

Marine Corps Is Studying Technologies to Reduce Fuel Consumption

The Marine Corps has taken steps to reduce its fuel usage by initiating research and development efforts to develop alternative power sources and improve fuel management. For example, it is testing the use of additional alternators in certain vehicles to provide onboard power capabilities, which could reduce the use of petroleum-based fuel and the number of generators needed on the battlefield. Another initiative involves providing hybrid power—by combining solar panel, generator, and battery energy sources—at remote sites to lessen fuel transportation demands to forward-deployed locations. The Marine Corps expects to begin testing this initiative in October 2008.

In addition, the Office of Naval Research is leading efforts for the Marine Corps to develop decision support tools that process and analyze data and improve fuel management in combat. Examples include sensors for fuel containers to measure the amount of remaining fuel and onboard vehicle sensors that automatically generate a requirement when additional fuel is needed.

DOD HAS NOT ESTABLISHED AN OVERARCHING ORGANIZATIONAL FRAMEWORK TO GUIDE AND OVERSEE MOBILITY ENERGY REDUCTION EFFORTS

While DOD and the military services have several efforts under way to reduce mobility energy demand, DOD lacks key elements of an overarching organizational framework to guide and oversee these efforts. As a result, DOD cannot be assured that its current efforts will be fully implemented and will significantly reduce its reliance on petroleum-based fuel. While DOD has identified energy as one of its transformational priorities, DOD's current approach to mobility energy lacks (1) top leadership, with a single executive-level OSD official—supported by an implementation team with dedicated resources and funding—who is accountable for mobility energy matters; (2) a comprehensive strategic plan for mobility energy; and (3) an effective mechanism to provide for communication and coordination of mobility energy efforts among OSD and the military services as well as leadership and accountability over each military service's efforts. In the absence of a framework for mobility energy that includes these elements, DOD has made limited progress in incorporating fuel efficiency as a consideration in its key business processes—which include developing requirements for and acquiring new weapons systems—and in implementing recommendations made in department-sponsored studies.

DOD's Current Management Approach to Mobility Energy Lacks Key Elements of an Overarching Organizational Framework

DOD's current approach to mobility energy is decentralized, with fuel oversight and management responsibilities diffused among several OSD and military service offices as well as working groups. More specifically, we found its approach lacks key elements of an overarching organizational framework, including a single executive-level OSD official—supported by an implementation team—who is accountable for mobility energy matters, a comprehensive strategic plan, and an effective mechanism for departmentwide communication and coordination. Our prior work on organizational transformations has found such a framework to be critical to successful transformation in both public and private organizations.[15] In addition, it is important to note that DOD has a history of creating organizational frameworks to address other crosscutting issues.

Responsibilities for Fuel Oversight and Management Are Diffused throughout Various DOD Offices and Working Groups

DOD's policies for energy management assign oversight and management responsibilities to several different offices without providing a single focal point with total visibility of, or accountability for, mobility energy reduction efforts across the department. Table 2 outlines various roles and responsibilities for fuel management and oversight.

Table 2. DOD Energy/Fuel Roles and Responsibilities

Office	Responsibilities
Under Secretary of Defense for Acquisition, Technology, and Logistics	Serve as the DOD senior energy official. Establish policies, grant waivers, and approve changes in the management of energy commodities.
Deputy Under Secretary of Defense (Logistics and Materiel Readiness)	Serve as the DOD central administrator for mobility energy policy with overall management responsibility for petroleum and other commodities.
Deputy Under Secretary of Defense (Installations and Environment)	Serve as the DOD central manager for facility energy policy on DOD installations.
Director, Defense Research and Engineering	Lead the DOD Energy Security Task Force.
DOD Comptroller	Establish financial policies and guidance for the management of energy commodities and related services, in coordination with the Under Secretary of Defense for Acquisition, Technology, and Logistics.
Defense Logistics Agency/Defense Energy Support Center	Serve as the executive agent for bulk petroleum and execute fuelrelated materiel management responsibilities.
Chairman, Joint Chiefs of Staff	Review operations plans and contingency plans to ensure that fuel requirements are addressed; identify fuel reporting requirements and other information for theater contingency requirements.
Combatant commanders' joint petroleum offices	Carry out combatant commander responsibilities for fuel distribution within a theater of operations.
Secretaries of the military departments	Operate the petroleum facilities under their cognizance, control fuel stocks in coordination with the Defense Logistics Agency, compute wartime fuel demands based on combatant commanders' operational and contingency plans.

Source: GAO analysis of DOD policies.

As table 2 shows, DOD policies do not assign responsibility for fuel reduction considerations—either singly or jointly—to any of the various offices involved in fuel management. While DOD directives designate the Under Secretary of Defense for Acquisition, Technology, and Logistics as the department's senior energy official, with responsibility for establishing policies, granting waivers, and approving changes in the management of energy commodities, including petroleum, the extent to which this official provides comprehensive guidance and oversight of fuel reduction efforts across the department is unclear.[16] Moreover, DOD has charged the Office of the Deputy Under Secretary of Defense (Logistics and Materiel Readiness) to serve as the DOD central administrator for mobility energy policy with overall management responsibility for petroleum and other commodities. We found that although this office plays an active role in maintaining DOD policy on energy supply issues and participates in other department-level fuel-related activities, its primary focus has not been on departmentwide fuel reduction efforts.

At the military service level, we found that the Air Force and the Army have established working groups to address fuel reduction and other energy issues. For example, the Air Force has established a senior focus group of high-level Air Force officials to address both mobility and facility energy issues. The senior focus group has created several working groups to address specific energy issues, such as aviation operations, acquisitions and technology, and synthetic fuels, as well as advisory groups on strategic communication, critical infrastructure protection, and financing. The Army also has established an energy working group to facilitate the discussion of energy issues across the service, including how to address rising fuel costs. The group meets each month to share information and identify issues across the Army. At the time of our review, the Army was in the process of establishing a senior steering group of high-level Army officials that would meet to discuss mutual energy concerns. While the Navy and Marine Corps have not established similar formal working groups, officials from both military services told us that they participate in internal meetings on fuel reduction issues.

DOD Has Not Designated a Single Executive-Level Official for Mobility Energy

While DOD has begun to increase management attention and has identified energy as a transformational priority, it has not designated a single executive-level OSD official—supported by an implementation team—who is accountable for mobility energy matters across the department. Our prior work has shown that top-level leadership and an implementation team with dedicated resources and funding are key elements of an overarching organizational framework. Furthermore, leadership must set the direction, pace, and tone and provide a clear, consistent rationale that brings everyone together behind a single mission.[17] The Under Secretary of Defense for Acquisition, Technology, and Logistics, as the senior DOD energy official, is responsible for management of energy commodities, but this individual also has a broad range of other responsibilities that include, among other things, matters relating to the DOD acquisition system, research and development, systems engineering, logistics, installation management, and business management modernization. Therefore, this individual's primary focus has not been on the management of mobility energy efforts. Moreover, from a broader perspective, the extent to which the Under Secretary of Defense for Acquisition, Technology, and Logistics has set a direction for the various OSD and military service offices involved in mobility energy is unclear.

In addition, DOD's Energy Security Task Force was formed in 2006 to address long-term departmental energy security requirements, such as DOD's reliance on fossil fuels, but we found that the task force has been unable to develop policy or provide guidance and oversight of mobility energy issues across the department. As indicated in its charter, the task force's integrated product team is required to develop a comprehensive DOD energy strategy and an implementation plan. Among other deliverables, the team's charter also requires it to define DOD's energy challenge, create a compendium of energy-related works, and perform a strategic assessment of energy. While the task force has taken steps to identify and monitor the progress of selected mobility energy reduction projects across the department, it has not yet completed an energy strategy or implementation plan, as well as other responsibilities. Furthermore, OSD officials told us that while the task force has briefed the Deputy Secretary of Defense's advisory group on its recommended projects, it does not have a "seat at the table" in departmental discussions at the Deputy Secretary of Defense level or at other executive levels, such as the Joint Requirements Oversight Council, the Defense Acquisition Boards, or the 3-Star Group within DOD's Planning, Programming, Budgeting, and Execution process.[18]

DOD also does not have an implementation team in place, with dedicated resources and funding, for mobility energy issues. For example, the officials who lead DOD's Energy Security Task Force's integrated product team do so as an extra responsibility outside of their normal work duties. Other DOD officials said that the task force provides a good forum for sharing energy ideas across the department, but lacks adequate staff to carry out specific actions. Furthermore, a task force participant told us that it can be difficult to find time to attend meetings while balancing other duties. The task force also does not receive any dedicated funding to pursue department-level energy priorities. Our prior work on the Government Performance and Results Act of 1993 (GPRA)[19] emphasizes the importance of relating funding to performance goals. The establishment of a dedicated funding mechanism for corrosion, for example, enabled DOD to fund high-priority corrosion reduction projects, which resulted in savings of more than $753 million during a 5-year period.[20] Without a long-term funding mechanism, DOD may not be able to ensure that mobility energy reduction efforts receive sustained funding over a period of years.

Moreover, DOD may not be well positioned to serve as a focal point on mobility energy within the department, with Congress, and with the Department of Energy or other interagency partners. During a military energy security forum held at the National Defense University in November 2007, representatives from various DOD offices presented energy as an area that is significant to a breadth of issues ranging from force protection to global stability to the security of DOD's critical infrastructure. They also noted that DOD has the potential to play multiple roles with respect to energy, including consumer, market leader, educator/motivator, oil infrastructure protector, and warfighter supporter. These concerns, coupled with an increased national and congressional interest in reducing fossil fuel dependence and exploring alternative energies, will likely necessitate an increased leadership focus on long-term energy issues, both within DOD and in its role as a stakeholder in interagency and national dialogues. The Energy Independence and Security Act of 2007,[21] for example, requires a variety of national-level actions, including that the President submit to Congress an annual report on the national energy security of the United States. It also requires DOD to examine energy and cost savings in nonbuilding applications, including an examination of savings associated with reducing the need for fuel delivery and logistical support. In addition, the John Warner National Defense Authorization Act for Fiscal Year 2007 directs DOD to improve the fuel efficiency of weapons platforms.[22]

DOD Has Not Yet Developed a Comprehensive Mobility Energy Strategic Plan

DOD has not yet developed a comprehensive strategic plan for mobility energy. Our prior work has found that strategic planning is a key element of an overarching organizational framework.[23] According to GPRA, key elements of a strategic plan include a comprehensive mission statement, goals and objectives, approaches or strategies to achieve those goals and objectives, and methods and timelines for evaluating progress. In addition, we have previously identified other elements that would enhance the usefulness of a strategic plan, including the development of outcome- oriented performance metrics and an alignment of activities, core processes, and resources to support mission-related outcomes.

DOD has taken some steps to lay the foundation for mobility energy strategic planning. According to OSD officials, DOD has begun to incorporate mobility energy issues into its *Guidance on the Development of the Force*, a department-level strategic planning document. In addition, the Office of the Deputy Assistant Secretary of Defense for Policy Planning, within the Office of the Under Secretary of Defense for Policy, is analyzing future energy concerns for the United States and the international security environment and highlighting their implications for the department. DOD officials said that the analysis is expected to provide information for consideration in the development of future strategic planning documents. We also observed that the DOD Energy Security Task Force has begun efforts to define goals that eventually may be incorporated into a DOD energy security strategic plan. OSD officials told us that the task force's intent is to complete this strategic plan by May 2008. However, current DOD strategic planning documents, such as the *National Military Strategy* and the most recent *Quadrennial Defense Review*,[24] do not address mobility energy reduction. Furthermore, until DOD fully develops and implements a comprehensive strategic plan for mobility energy, it cannot be certain that mobility energy reduction efforts align with the department's energy mission or strategic goals to ensure that they are appropriately prioritized, or know whether critical gaps or duplication of efforts exist.

DOD Does Not Have an Effective Mechanism for Communication and Cross- Service Coordination of Mobility Energy Reduction Efforts

DOD does not have an effective mechanism to facilitate communication and coordination of mobility energy reduction efforts among OSD and the military services. Our prior work has shown that a communication strategy involves creating shared expectations and reporting related progress.[25] While DOD's Energy Security Task Force aims to identify key players within the energy field, its current structure does not ensure departmentwide communication of fuel reduction efforts, particularly among the military services, which are responsible for most of these efforts. More specifically, during our observation of a task force monthly meeting, we found that although this venue provides for some sharing of information, the generally less than 2 hours allotted for each monthly meeting does not allow for effective coverage of the spectrum of DOD's mobility energy issues. Moreover, we noted that although the task force's senior steering group includes, among others, the service under secretaries and assistant secretaries; the Director, Defense Research and Engineering; and several principal deputy under secretaries of defense, it only meets two to three times a year. Furthermore, with the exception of the Air Force, none of the other military service members on the senior steering group have primary responsibility for mobility energy reduction efforts within their services. Without executive-level focal points, the military services may not be well positioned to effectively coordinate on mobility energy reduction efforts across the department or provide leadership or accountability for efforts within their services.

In addition, we found a lack of cross-service coordination concerning mobility energy reduction initiatives. Army officials told us that they were unaware of Navy research on fuel reduction metrics, while Air Force officials said that they do not routinely discuss aviation fuel reduction initiatives with their Army counterparts, even though both military services are concerned about aircraft fuel consumption. OSD officials said that while several separate groups are making efforts to reduce fuel consumption, the efforts are often not shared or integrated. Moreover, OSD officials told us that DOD generally lacks incentives to reward the military services for reducing fuel consumption and faces challenges in addressing departmental cultural barriers—such as the traditional view that fuel is simply a commodity and that energy efficiency is not an important consideration for warfighting. Without an effective mechanism to facilitate communication of mobility energy reduction efforts between OSD and the military services, DOD cannot be certain that these efforts are effectively coordinated throughout the department or consistent with DOD's energy priorities and goals. On a broader level, DOD may not be well positioned to respond to congressional or other agencies' requests for information on mobility energy.

DOD Has a History of Creating Organizational Frameworks to Address Other Crosscutting Issues

Many OSD, military service, and other DOD officials with whom we spoke expressed the need for an overarching organizational framework to address mobility energy throughout the department. Some officials from OSD suggested that an ideal organizational framework would bring together the various offices within OSD and the military services involved in fuel reduction efforts and establish business practices, analytic methods, and technology investments that take into account strategic risks associated with energy. Some military service officials acknowledged that departmental oversight is needed but told us that they fear such oversight might take resources away from their own mobility energy reduction initiatives. Similarly, some OSD officials said they are concerned that establishing a permanent mobility energy office or similar framework could impose additional bureaucratic layers and slow progress on mobility energy reduction initiatives.

We noted that DOD has established new organizational frameworks to address other crosscutting issues, such as business systems modernization, corrosion control and prevention, contractors on the battlefield, and the defeat of improvised explosive devices. While we did not evaluate the strengths or weaknesses of these organizational frameworks as part of this review, they nonetheless provide DOD examples to consider in determining how best to establish an overarching organizational framework for mobility energy. For example, the Business Transformation Agency, which addresses business systems modernization, involves top DOD leadership by operating under the Office of the Under Secretary of Defense for Acquisition, Technology, and Logistics but reporting directly to the Deputy Under Secretary of Defense for Business Transformation. DOD has also created a management framework to oversee facility energy, which accounts for about 25 percent of the department's energy use. Specifically, it has designated a senior agency official, the Deputy Under Secretary of Defense for Installations and Environment, with the responsibilities for meeting federal mandates regarding energy reduction at installations. The department has also created a working group charged with implementing the mandates.

In addition, DOD established an Energy Policy Council in 1985 to provide coordinated review of DOD energy policies, issues, systems, and programs.[26] In the instruction outlining the requirements of this council, DOD assigned responsibilities to various departmental offices and designated the then Deputy Assistant Secretary of Defense (Logistics and Materiel Management) as council chair. DOD also called for clearly identified focal points to address energy matters within each military department. When we asked about the status of the council, OSD officials said that they did not believe it still existed. This now-defunct Energy Policy Council could also serve as an example of an organizational framework for mobility energy that provides for sharing of information among the military services.

Absence of an Overarching Organizational Framework Does Not Position DOD to Effectively Address Mobility Energy

In the absence of an overarching organizational framework, DOD is not well positioned to fully incorporate fuel efficiency considerations into its key business processes or to fully implement recommendations from DOD-sponsored studies on fuel reduction.

DOD Has Not Yet Fully Incorporated Fuel Efficiency Considerations into Its Key Business Processes

DOD has not yet fully incorporated fuel efficiency considerations into key departmental business processes, such as its requirements development and acquisition processes for new weapons platforms and other mobile defense systems. DOD's process to develop requirements, known as the Joint Capabilities Integration and Development System, is a multistep process that involves identifying what military capabilities the department needs to accomplish its tasks. Once the capabilities are identified, DOD's acquisition process produces equipment that can meet those requirements. DOD-sponsored studies on fuel reduction, such as the 2007 LMI report, note that the requirements development and acquisition processes provide opportunities for DOD to consider energy efficiencies while considering capabilities. Moreover, the 2001 Defense Science Board report noted that fuel efficiency benefits are not currently valued or emphasized in DOD's requirements development and acquisition processes. While DOD has recently begun to take some steps to integrate fuel considerations into these processes, these considerations are not factored in a systematic manner and cannot be fully applied.

For example, DOD's requirements development process does not systematically include energy efficiency considerations, and the capability gap assessments associated with the process do not include fuel-related logistics, thus leaving these types of issues to be resolved after systems are fielded. As described earlier, in May 2007, the Joint Staff established an energy efficiency key performance parameter that would require fuel considerations during capabilities development. However, because DOD has not developed a methodology to determine how best to employ the energy efficiency key performance parameter, implementation of this key performance parameter remains uncertain.

DOD has also taken steps to inform its acquisition process with its pilot program to determine the fully burdened cost of fuel for three mobile defense systems. While the pilot program represents a step toward providing visibility over the total logistics costs associated with delivered fuel and DOD has set a fall 2008 deadline to issue guidance for applying the fully burdened cost of fuel in acquisition programs, DOD has not yet developed an approach for determining how it would incorporate this information into its acquisition decision-making process. Moreover, the 2008 Defense Science Board report presented some concerns about how fully burdened costs are being calculated. Specifically, the report cited a concern that the analysis focused on peacetime costs and did not adequately consider wartime costs, even though the fully burdened cost analysis is intended to be a wartime capability planning factor. Until the pilot program is completed and the results are assessed, DOD is not in a position to apply a fully burdened cost analysis to its acquisition process. Thus, the department is unable to promote greater visibility over its acquisition decisions or more fully consider the operational and cost consequences of the fuel burden on the logistics infrastructure.

Other key DOD business processes, such as those that address repair, recapitalization, and replacement of mobile defense systems also present opportunities to incorporate fuel efficiency measures during system upgrades. However, OSD officials told us that the department generally makes decisions about system upgrades without regard to fuel efficiency, including the fully burdened cost, in part because such decisions require greater up-front costs. Although DOD recognizes that by reducing energy demand it can provide its forces greater flexibility and reduce their dependence on the logistics infrastructure, some OSD officials told us that DOD's budget process promotes a short-term outlook and does not encourage the purchase of fuel-efficient systems or upgrades that may initially cost more but could reduce life cycle and logistics costs over the long term. Moreover, the 2008 Defense Science Board report noted that DOD's lack of tools to assess the operational and economic benefits of fuel efficiency technologies is a major reason why DOD underinvests in the development and deployment of these technologies. In addition, OSD officials told us that DOD does not systematically assess how making fuel efficiency upgrades to systems would affect other logistics issues—for example, how reducing the weight of an Army vehicle would affect the amount of fuel the Air Force transports to the battlefield for that vehicle. Such assessments, they said, may reveal further enhancements in warfighting capabilities.

DOD Has Been Slow to Implement Recommendations from Department-Sponsored Studies on Fuel Reduction

In the absence of an overarching organizational framework, DOD has made limited progress in implementing recommendations from department-sponsored studies by organizations such as the Defense Science Board, The JASONs, and LMI that have urged an expansion of efforts to reduce dependency on petroleum-based fuel. These studies confirmed that, for many reasons, continued heavy reliance on petroleum-based fuel poses a significant problem for DOD. For example, LMI reported that DOD's increasing fuel demand furthers the nation's reliance on foreign energy sources and limits the department's ability to establish a more mobile and agile force. The studies found a need to focus more DOD management attention on mobility energy matters and recommended actions aimed at, among other things, improving the fuel efficiency of weapons platforms, eliminating institutional barriers that bear upon the department's decisions regarding fuel efficiency, and developing a longterm mobility energy strategy that would lead to reduced consumption of petroleum-based fuel.

DOD has not taken a formal position on these recommendations, and implementation, in some cases, would require significant changes throughout the department that could generate institutional resistance. One study, for example, called for creating a unified energy governance structure in order to alter DOD's "energy culture." During our review, we found that DOD had taken some steps toward implementing some of the recommendations, such as initiating a pilot program for determining the fully burdened cost of delivered fuel and adding a requirement for an energy efficiency key performance parameter in its Joint Staff policy manual. However, other recommendations, such as establishing a governance structure for mobility energy, have not been implemented (see app. II for our summary of the recommendations in DOD-sponsored studies and the actions DOD has taken on those recommendations). The 2008 Defense Science Board report noted that the recommendations made by the 2001 Defense Science Board report are still open and remain viable. An overarching organizational framework could better position DOD to address these and other fuel reduction recommendations in a more timely and effective manner. Moreover, a framework for mobility energy could provide greater assurance that DOD's efforts to reduce its reliance on petroleum-based fuel will succeed without degrading its operational capabilities and that DOD is better positioned to address future mobility energy challenges.

CONCLUSIONS

DOD continues to face rapidly increasing fuel costs and high fuel requirements that have placed a significant logistics burden on its forces. In light of these and other challenges associated with mobility energy, DOD has begun to increase its management attention on reducing its reliance on petroleum-based fuel. Increased national focus on the United States' dependence on foreign oil, projected increases in the worldwide demand for oil, and uncertainties about world oil supplies will likely require DOD to further increase its focus on long-term energy issues, both within the department and as a stakeholder in interagency and national dialogues. However, DOD will have difficulty addressing mobility energy challenges in the absence of an overarching organizational framework. Without such a framework, DOD is not well positioned to effectively guide and oversee mobility energy reduction efforts from a departmentwide perspective to ensure that efforts are appropriately prioritized; identify critical gaps or duplication of efforts; and address long-term, large-scale energy issues. In particular, no individual at the executive level within OSD has been designated to be accountable for mobility energy and set the direction, pace, and tone to reduce mobility energy demand across the department. Other elements of an overarching organizational framework include a comprehensive strategic plan and executive-level focal points at the military services to provide for effective coordination. In addition, until DOD takes steps to further incorporate energy efficiency considerations into its business processes, the department is unable to promote greater visibility in its decision making or fully consider the effects of fuel on the logistics infrastructure. With a mobility energy overarching organizational framework in place, DOD would be better positioned to reduce its significant reliance on petroleum-based fuel and to address the energy challenges of the 21st century.

RECOMMENDATIONS FOR EXECUTIVE ACTION

To improve DOD's ability to guide and oversee mobility energy reduction efforts, we recommend that the Secretary of Defense direct the Deputy Secretary of Defense to establish an overarching organizational framework by taking the following three actions:

- Designate an executive-level OSD official who is accountable for mobility energy matters and sets the direction, pace, and tone to reduce mobility energy demand across the department; improve business processes to incorporate energy efficiency considerations as a factor in DOD decision making; coordinate on energy issues with facility energy officials; act as DOD's focal point in interagency deliberations about national energy concerns; and lead the department's potential transition from petroleum- based fuel to alternative fuel sources. This official should be supported by an implementation team with dedicated resources and funding.
- Direct the executive-level mobility energy official to lead the development and implementation of a comprehensive departmentwide strategic plan for mobility energy. At a minimum, this strategic plan should set forth mobility energy goals and objectives, time frames for implementation, and performance metrics to track and evaluate progress.
- Ensure that OSD takes the following steps to fully incorporate energy efficiency considerations into DOD's requirements development and acquisition processes:
- Develop a methodology to enable the full implementation of an energy efficiency key performance parameter in DOD's requirements development process.
- As part of its efforts to complete DOD's fully burdened cost of fuel pilot program, develop an approach for incorporating this cost information into the acquisition decision making process.

Furthermore, to establish effective communication and coordination among the executive-level OSD mobility energy official and the military services, we recommend that the Secretary of Defense direct the Secretaries of the Army, Navy, and Air Force and the Commandant of the Marine Corps to designate an executive-level official within each of their military services to act as a focal point on departmentwide mobility energy efforts as well as provide leadership and accountability over their own efforts.

AGENCY COMMENTS AND OUR EVALUATION

In its written comments on a draft of this chapter, DOD partially concurred with all of our recommendations. Based on DOD's comments to our draft chapter, we made minor modifications to our chapter, including our first recommendation. Technical comments were provided separately and incorporated as appropriate. The department's written comments are reprinted in appendix III.

In response to our recommendation that the Secretary of Defense direct the Deputy Secretary of Defense to designate an executive-level OSD official who is accountable for mobility energy matters across the department, DOD acknowledged that there is a need to view and manage its energy challenges in a new, more systematic manner. DOD's response stated that DOD Directive 5134.01 (Dec. 9, 2005) provides the Under Secretary of Defense for Acquisition, Technology, and Logistics oversight and policy-making authority on DOD energy matters. However, it is clear from our review, including discussions with department officials, that neither the Under Secretary nor any official from this office is providing comprehensive oversight and policy guidance for mobility energy across the department. Instead, we found that DOD's current approach to mobility energy is decentralized, with fuel oversight and management responsibilities diffused among several OSD and military service offices (see table 2 of this chapter) as well as working groups. DOD does not assign responsibility for fuel reduction considerations—either singly or jointly— to any of the various offices involved in fuel management. DOD's response stated that its authorities and responsibilities are consistent with those used for overseeing other significant crosscutting issues. However, as we noted in our chapter, DOD has established new organizational frameworks to address other crosscutting issues, such as business systems modernization, corrosion control and prevention, contractors on the battlefield, and the defeat of improvised explosive devices. Moreover, DOD has established a focal point for facility energy, the Deputy Under Secretary of Defense for Installations and Environment, within the Office of the Under Secretary of Defense for Acquisition, Technology, and Logistics, even though facility energy accounts for about 25 percent of DOD's total energy consumption. Mobility energy accounts for about three-fourths of its total energy consumption, but there is not an equivalent focal point. Key energy issues—including rising fuel costs, worldwide energy demand, and the high fuel burden during operations— underscore the importance of energy to DOD and will likely require sustained top leadership attention. DOD stated that significant mobility energy efforts are currently under way that will provide for better management of mobility energy. While we acknowledge that DOD has begun to increase management attention on mobility energy issues by creating the DOD Energy Security Task Force, the department does not have an implementation team, with dedicated resources and funding, for mobility energy issues. As we noted in our chapter, the task force's current structure does not ensure departmentwide communication of fuel- reduction efforts, particularly among the military services, which are responsible for most of these efforts. Based on DOD's response to our first recommendation, we made minor modifications to the recommendation to emphasize that DOD should designate an executive-level OSD mobility energy official—supported by an implementation team—who is accountable for mobility energy matters and who sets the direction, pace, and tone to reduce mobility energy demand across the department. This official should also improve business practices to incorporate energy considerations as a factor in DOD decision making; coordinate on energy issues with facility energy officials; act as DOD's focal point in interagency deliberations about national energy concerns; and lead the department's potential transition from petroleum-based fuel to alternative fuel sources. Without such an official to provide this leadership, DOD is not well positioned to address mobility energy challenges.

In response to our recommendation that the Secretary of Defense direct the Deputy Secretary of Defense to direct the executive-level mobility energy official to lead the development and implementation of a comprehensive departmentwide strategic plan for mobility energy, DOD indicated that the Under Secretary of Defense for Acquisition, Technology, and Logistics is overseeing the development of a DOD energy security strategic plan which will be reported to the Deputy's Advisory Working Group in May 2008. We believe that this is a step in the right direction. As we noted in this chapter, until DOD fully develops and implements a comprehensive strategic plan for mobility energy—that sets forth mobility energy goals and objectives, time frames for implementation, and performance metrics to track and evaluate progress—DOD will not be able to ensure that mobility energy reduction efforts align with the department's energy mission or strategic goals to ensure that they are appropriately prioritized, or to know whether critical gaps or duplication of efforts exist.

In response to our recommendation that the Deputy Secretary of Defense ensure that OSD takes steps to fully incorporate energy efficiency considerations into DOD's requirements development process by developing a methodology to enable the full implementation of an energy efficiency key performance parameter, DOD stated that it plans to address how and when it will implement such a methodology in its forthcoming DOD energy security strategic plan. However, this plan does not yet exist.

Because DOD is linking the development of a methodology for an energy efficiency key performance parameter to this plan, the implementation of the key performance parameter remains uncertain. Thus DOD cannot ensure that energy efficiency considerations are factored into its requirements development process in a systematic manner. In addition, in response to our recommendation that DOD develop an approach for incorporating the information from its fully burdened cost of fuel pilot program into its acquisition process, DOD stated that it is developing a plan on how best to assess fuel efficiency relative to the costs and operational capabilities of its weapons systems. Again, until this plan is completed, DOD is not in a position to apply a fully burdened cost analysis to its acquisition process. Thus, the department is unable to promote greater visibility over its acquisition decisions or more fully consider the operational and cost consequences of the fuel burden on the logistics infrastructure.

In response to our recommendation that the Secretary of Defense direct the Secretaries of the Army, Navy, and Air Force and the Commandant of the Marine Corps to designate an executive-level official within each of their military services to act as a focal point on departmentwide mobility energy efforts as well as provide leadership and accountability over their own efforts, DOD stated that it will address this issue after it has briefed the DOD energy security strategic plan to DOD senior leaders in May 2008. However, as we noted in this chapter, a lack of cross-service coordination concerning mobility energy reduction initiatives currently exists. By waiting to address this issue, the department cannot be certain that the mobility energy efforts of the military services are consistent with the department's energy priorities and goals. Designating executive-level military service focal points would provide improved leadership and accountability over their own efforts as well as increased coordination across the department.

William M. Solis
Director
Defense Capabilities and Management

APPENDIX I: SCOPE AND METHODOLOGY

To address our objectives, we focused our work on the Department of Defense's (DOD) mobility energy issues related to fuel demand for operations. We did not address supply issues, fuel for nontactical vehicles, or DOD facility energy management, except to briefly describe the organizational structure DOD employs to manage energy issues at its fixed installations.

To identify key departmental and military service efforts that have been undertaken to reduce demand for mobility energy, we obtained and reviewed documentation from the Office of the Secretary of Defense (OSD), the Joint Staff, and the military services on their key mobility energy reduction efforts. These documents included briefings, policies, directives, military service studies, and associated paperwork on the specific efforts. We also interviewed cognizant departmental and military service officials who identified and provided the documentation for key efforts. At the department level, we spoke with officials involved with the DOD Energy Security Task Force, including members of the integrated product team and working groups, to obtain information about the task force's goals, accomplishments, and challenges as well as the specific service mobility energy initiatives it has chosen to monitor. We also interviewed OSD and Joint Staff officials to obtain information on their efforts to incorporate energy efficiency considerations into DOD's requirements development and acquisition processes. At the military service level, we interviewed officials to determine how each military service is approaching its specific mobility energy reduction efforts, its progress to date, and what challenges it faces in reducing mobility energy demand. We did not validate the cost estimates provided by the services for their initiatives. To obtain a broad perspective of the energy issues, we attended two defense-related conferences that focused on national security energy concerns and their potential implications for DOD.

To assess the extent to which DOD has established an overarching organizational framework to guide and oversee mobility energy efforts, we reviewed and analyzed DOD documentation, such as policies and directives, DOD-sponsored fuel-related studies, and legislation, and interviewed officials from OSD, the Joint Staff, and the military services. In doing so, we examined DOD's key business processes, such as its requirements development and acquisition processes, and determined the extent to which fuel efficiency is systematically considered in these processes. We also identified key elements of an overarching organizational framework based on our prior work and the Government Performance and Results Act of 1993 to determine the extent to which DOD's current structure incorporated or lacked these key elements. We interviewed officials at OSD and the military services to obtain their perspectives on DOD's current approach to mobility energy, including the extent to which the DOD Energy Security Task Force is developing policy and providing guidance and oversight of mobility energy issues across DOD. We also attended a meeting of the Energy Security Task Force's integrated product team to observe the format, content, participants, and dialogue of a typical meeting. In addition, we asked the officials about what benefits and consequences they saw with the existing department- level involvement (or lack thereof) in mobility energy issues. We also identified management frameworks DOD has created to address other crosscutting issues, such as business systems modernization, corrosion control and prevention, contractors on the battlefield, the defeat of improvised explosive devices, and facility energy. We did not evaluate the strengths or weaknesses of these organizational frameworks or their specific applicability to mobility energy. We also reviewed DOD-sponsored studies published since 2000 on reducing fuel demand in DOD's mobile defense systems, focusing on studies that made recommendations specific to departmentwide mobility energy issues. After an initial literature search and discussions with DOD officials and other researchers, independent of DOD, we ultimately selected four studies to include in our review. We interviewed coauthors from each of these studies to gain a better understanding of their objectives, scopes, and methodologies and their perspectives on the issues covered in their reports as well as other department-level mobility energy concerns. Two team members consolidated the recommendations related to mobility energy from these studies and analyzed them for similarities. They combined those that were similar, rephrased the wording while keeping the intent, and categorized the recommendations into common themes. Through their review of documentation and interviews with DOD officials, they then summarized the actions taken on each of the recommendations. A third team member independently reviewed the results, and discussed any discrepancies with the other team members to reach agreement on the appropriate themes and actions taken.

We coordinated our work at the following DOD offices:

Office of the Secretary of Defense

- Office of the Under Secretary of Defense for Acquisition, Technology, and Logistics
- Systems Engineering and Developmental Test and Evaluation
- Office of the Director, Defense Research and Engineering
- Office of the Deputy Under Secretary of Defense, Logistics and Materiel Readiness
- Office of the Under Secretary of Defense Comptroller/Chief Financial Officer
- Office of the Under Secretary of Defense for Policy

- Office of the Deputy Assistant Secretary of Defense for Policy Planning
- Office of the Deputy Assistant Secretary of Defense for Forces Transformation and Resources
- Director of Program Analysis and Evaluation

Chairman, Joint Chiefs of Staff

- Logistics (J4)
- Operational Plans and Joint Force Development (J7)
- Force Structure, Resources, and Assessment (J8)

Department of the Army

- Army Deputy Chief of Staff (G4)
- Assistant Secretary of the Army for Acquisition, Logistics, and Technology
- U.S. Army Combined Arms Support Command
- Army Rapid Equipping Force

Department of the Navy

- Office of the Chief of Naval Operations
- Naval Sea Systems Command
- Office of Naval Research
- Headquarters, Marine Corps

Department of the Air Force

- Office of the Deputy Assistant Secretary of the Air Force for Environment, Safety and Occupational Health
- Logistics, Installations and Mission Support (A4/7)
- Strategic Plans and Programs (A8)
- Conduct Air, Space, and Cyber Operations

Other DOD Components

- United States Joint Forces Command
- Defense Logistics Agency/Defense Energy Support Center

We conducted our review from September 2007 through March 2008 in accordance with generally accepted government auditing standards.

APPENDIX II: SELECT RECOMMENDATIONS FROM DOD-SPONSORED STUDIES ON MOBILITY ENERGY REDUCTION

Over the past 7 years, DOD has commissioned several studies to explore ways to reduce its fuel consumption. We reviewed recommendations applicable to mobility energy in the following three DOD-sponsored studies:

- Defense Science Board, More Capable Warfighting Through Reduced Fuel Burden, January 2001
- The JASONs/The MITRE Corporation, Reducing DOD Fossil-Fuel Dependence, September 2006
- LMI, Transforming the Way DOD Looks at Energy: An Approach to Establishing an Energy Strategy, April 2007

We also reviewed the recommendations from the 2008 Defense Science Board report on DOD's energy strategy. However, we did not include those recommendations in our analysis because the report was issued in February 2008, and the department could not be expected to have taken action on the recommendations at the time we issued this chapter.

We summarized the recommendations, grouped them into common topics, and obtained information on DOD actions taken on each of them. Table 3 presents a summary of our analysis.

Table 3. Actions DOD Has Taken to Address Selected Recommendations from DOD-Sponsored Studies on Mobility Energy Reduction

Topic/recommendation	Source	DOD action taken
Technology upgrades/system redesigns		
Consider fuel efficiency when making science and technology and system design investments.	*More Capable Warfighting Through Reduced Fuel Burden* by the Defense Science Board	The Office of the Under Secretary of Defense for Acquisition, Technology, and Logistics initiated a pilot program to assess the fully burdened cost of fuel in three mobile defense systems, and the Joint Staff established an energy efficiency key performance parameter. Full implementation of both efforts could provide insight on how to include energy considerations in system design. Moreover, according to OSD officials, DOD's acquisition process currently undervalues energy efficiencies. Thus, investments may not be adequate.
Upgrade the engines of the M1- Abrams tank, the B-52 bomber, and other applicable systems with modern fuel-efficient engine technology.	*Reducing DOD Fossil-Fuel Dependence* by The JASONs/The MITRE Corporation	DOD provided recommendations to the military services for consideration and implementation.
Reduce weight of armored and tactical vehicles, with modern vehicle designs, structures, and materials.	*Reducing DOD Fossil-Fuel Dependence* by The JASONs/The MITRE Corporation	The Ground Fuel-Efficient Vehicle Demonstrator, an initiative funded in fiscal year 2008, will examine fuelefficient technologies and equipment for ground vehicle programs.
Develop a DOD-wide system to track vehicle and fuel use patterns. Use the data to develop DOD-wide fuel efficiency metrics in decisions to upgrade system designs.	*Reducing DOD Fossil-Fuel Dependence* by The JASONs/The MITRE Corporation	The military services have some systems or reporting mechanisms for capturing fuel consumption. However, as a department, DOD has not developed metrics for fuel consumption.

Table 3. (Continued).

Topic/recommendation	Source	DOD action taken
Assess options for expanding the use of unmanned vehicles by considering more fuel-efficient designs.	*Reducing DOD Fossil-Fuel Dependence* by The JASONs/The MITRE Corporation	In fiscal year 2007, the Office of the Director, Defense Research and Engineering, initiated the Long Endurance Unmanned Aerial Vehicle program, which is intended to increase the amount of time an unmanned aerial vehicle could stay in the air without refueling.
Strategic planning		
Develop tools to track and account for the total costs of fuel, including delivery and logistics costs.	*More Capable Warfighting Through Reduced Fuel Burden* by the Defense Science Board *Reducing DOD Fossil-Fuel Dependence* by The JASONs/The MITRE Corporation	The Office of the Under Secretary of Defense for Acquisition, Technology, and Logistics initiated a pilot program to assess the fully burdened cost of fuel in three mobile defense systems. However, the program has not been implemented for all systems.
Increase simulator use.	*Reducing DOD Fossil-Fuel Dependence* by The JASONs/The MITRE Corporation	In August 2007, the Office of the Under Secretary of Defense for Acquisition, Technology, and Logistics directed a study to assess whether the increased use of simulators could substitute for live training without degrading operational capability. An implementation plan has been drafted and a working group has been established.
Integrate the military services' fuel requirements into logistics war games and analytic tools instead of assuming that fuel supplies will be adequate.	*More Capable Warfighting Through Reduced Fuel Burden* by the Defense Science Board *Reducing DOD Fossil-Fuel Dependence* by The JASONs/The MITRE Corporation	DOD incorporated energy considerations into the annual Unified Engagement exercise, but it is not a standard consideration in DOD's war-gaming processes.

Topic/recommendation	Source	DOD action taken
Include fuel efficiency as a key performance parameter in DOD's requirements process.	*More Capable Warfighting Through Reduced Fuel Burden* by the Defense Science Board	The Joint Staff has established an energy efficiency key performance parameter. However, DOD has not developed a methodology to fully implement this requirement.
Engage in long-term planning for future fuel sources, production, and use.	*Reducing DOD Fossil-Fuel Dependence* by The JASONs/The MITRE Corporation	DOD has not developed a comprehensive strategic plan for mobility energy.
Increase the use of commercial aviation fuels; consider the local production of military fuels from commercial aviation fuels outside of the United States.	*Reducing DOD Fossil-Fuel Dependence* by The JASONs/The MITRE Corporation	The Air Force has conducted one study examining this issue, and the Defense Energy Support Center is proposing an additional study to better understand the logistical impacts of using commercial aviation fuels.
Incorporate energy considerations (use and logistics requirements) into DOD's key corporate decision making processes: strategic planning; analytic agenda; joint concept and joint capability development; acquisition; and planning, programming, budgeting, and execution.	*Transforming the Way DOD Looks at Energy: An Approach to Establishing an Energy Strategy* by LMI	While DOD has made some efforts to address energy efficiency in its requirements development and acquisition processes, these efforts are in the early stages.
Leadership and oversight		
Adopt a new framework to promote energy efficiency, including alternate energy sources, to those areas (1) consuming the most fuel aviation forces), (2) requiring the most logistics support, or 3) having the most negative effect on the warfighter.	*Transforming the Way DOD Looks at Energy: An Approach to Establishing an Energy Strategy* by LMI	The DOD Energy Security Task Force has selected military service initiatives to monitor that address energy efficiency in selected areas, but DOD has not developed an overarching organizational framework for mobility energy.

Table 3. (Continued).

Topic/recommendation	Source	DOD action taken
Provide leadership that incentivizes fuel efficiency throughout DOD.	*More Capable Warfighting Through Reduced Fuel Burden* by the Defense Science Board	OSD officials told us that DOD generally lacks incentives to reward the military services for reducing fuel consumption and faces challenges in addressing departmental cultural barriers—such as the traditional view that fuel is simply a commodity and that energy efficiency is not important to warfighting.
Establish a governance structure with policy and resource oversight to focus DOD's energy efforts.	*Transforming the Way DOD Looks at Energy: An Approach to Establishing an Energy Strategy* by LMI	DOD has not established an overarching organizational framework to provide oversight for mobility energy.

Source: GAO analysis of DOD information.

APPENDIX III: COMMENTS FROM THE DEPARTMENT OF DEFENSE

OFFICE OF THE UNDER SECRETARY OF DEFENSE
3000 DEFENSE PENTAGON
WASHINGTON, DC 20301-3000

ACQUISITION, TECHNOLOGY AND LOGISTICS

FEB 2 0 2008

Mr. William M. Solis
Director, Defense Capabilities and Management
U.S. Government Accountability Office
441 G Street, N.W.
Washington, DC 20548

Dear Mr. Solis:

This is the Department of Defense (DoD) response to the GAO draft report, GAO-08-426, "DEFENSE MANAGEMENT: Overarching Organizational Framework Needed to Guide and Oversee Energy Reduction Efforts for Military Operations," dated February 5, 2008 (GAO Code 351082). Detailed comments on the report recommendations are enclosed.

The Department of Defense partially concurs on the GAO's findings and recommendations. In principle, the Department agrees there is a need to view and manage DoD energy supply and demand challenges in a new, more systemic manner. However, current DoD Directives are clear on where the authority and responsibility lie on the issues surrounding mobility energy. DoD Directive 5134.01 (Dec 9, 2005) puts oversight and policymaking authority on the DoD energy matters the GAO raised under the authority of the Under Secretary of Defense for Acquisition, Technology and Logistics. The Under Secretary's staff is leading a Department-wide effort to develop a comprehensive DoD Energy Security Strategic Plan ready to be delivered to senior DoD leaders this May. Further, the Department has dedicated resources and senior-level attention to implementing an Energy Efficiency Key Performance Parameter, underpinned by maturing and applying the Fully Burdened Cost of Fuel construct to its requirements and acquisition tradespace analysis and decision-making.

The Department appreciates this opportunity to comment on this draft report.

Sincerely,

Kristen J. Baldwin
Acting Director
Systems and Software Engineering

GAO DRAFT REPORT – DATED FEBRUARY 5, 2008
GAO CODE 351082 /GAO-08-426

"DEFENSE MANAGEMENT: Overarching Organizational Framework Needed to Guide and Oversee Energy Reduction Efforts for Military Operations"

DEPARTMENT OF DEFENSE COMMENTS TO THE RECOMMENDATIONS

RECOMMENDATION 1: The GAO recommends that the Secretary of Defense direct the Deputy Secretary of Defense to establish an overarching organizational framework by designating an executive-level OSD official with the responsibility and authority to guide and oversee efforts and develop policy, among other things, to reduce mobility energy demand across the Department; improve business processes to incorporate energy considerations as a factor in DoD decision making; coordinate on energy issues with facility energy officials; act as DoD's focal point in interagency deliberations about national energy concerns; and lead the Department's potential transition from petroleum-based fuel sources to alternative fuel sources. This official should be supported by an implementation team with dedicated resources and funding.

DOD RESPONSE: Partially concur. In principle, the Department concurs that there is a need to view and manage DoD energy supply and demand challenges in a new, more systematic manner. The direct cost of mobility fuels, the indirect (yet substantial) cost of our fuel delivery logistics, the operational vulnerability of our logistics forces to insurgent forces and other threats, and our prioritization of science and technology (S&T) and acquisition investments seeking greater energy efficiency are issues of increasing importance. However, significant efforts are currently underway that will provide the Department's senior leaders new analytic tools, business process changes and S&T energy efficiency investments recommendations for better managing DoD mobility energy across the enterprise.

These efforts, particularly those of the DoD Energy Security Task Force (ESTF), which is now developing a comprehensive Energy Security Strategic Plan for the DoD, will be delivered to the Deputy Secretary's Advisory Working Group in May 2008. The work of the ESTF will provide new insights and recommendations to DoD's decision-makers for the governance of mobility energy issues across the Department. This work is informed by operational experience in the field, including from Operations Enduring Freedom and Iraqi Freedom, as well as the independent analysis and perspective of the 2008 Defense Science Board (DSB) Task Force on Energy Security. This task force, which contained a strong balance of seasoned former national security leaders, including a former Secretary of Defense and of Energy, a former CIA Director, and retired senior flag officers, as well as professional energy and energy efficiency experts, took a hard look at DoD risks and possible solutions related to energy. The 2008 DSB Task Force on Energy Security was a consensus document and identified the many and difficult challenges the Department must now take on. In this case, consensus was not the result of watering down the findings, but rather, solid agreement on the risks and underappreciated factors that go into planning for DoD energy supply and demand.

Further, addressing the governance issues the GAO report raises, current DoD Directives are clear on where the authority and responsibility lie on the issues surrounding mobility energy. DoD Directive 5134.01 (Dec 9, 2005) puts oversight and policymaking authority on DoD energy matters under the authority of the Under Secretary of Defense for Acquisition, Technology and Logistics, particularly in sections 3, 3.24, 3.26, E2.1.1.7, E2.1.1.8, E2.1.1.28, E2.1.1.29 and E2.1.1.30. These authorities and responsibilities are consistent with those used for overseeing many other significant cross-cutting DoD capability development and management issues.

The Under Secretary of Defense (AT&L) oversees three senior executives with policy-making and oversight roles on different aspects of mobility energy. The Deputy Under Secretary of Defense for Acquisition and Technology (DUSD (A&T)) oversees the system acquisition process, and has the authority to provide guidance, as needed, to the Components on how the acquisition tradespace is considered, both from a platform and capability perspective. The DUSD (A&T) organization is working within the OSD staff, Joint Staff and Service force planning and requirements communities to ensure fuel demand, fuel logistics and all related matters are given more appropriate consideration in the acquisition tradespace to reduce energy demand. This includes maturing a construct known as the "Fully Burdened Cost of Fuel (FBCF)" and applying the FBCF principles and methodology to the current Energy Efficiency Key Performance Parameter (KPP) in the Joint Capabilities Integration and Development System. This is the process by which materiel and non-materiel solutions to documented capability gaps are identified and framed for the acquisition and other communities to solve. This work will serve as a basis for the Department's senior leaders to assess DoD mobility fuel demand from a portfolio perspective (force application, force protection, etc.) and not just from a platform perspective. This approach is consistent with and supportive of the Deputy Secretary's management agenda. Taking a portfolio perspective will better inform risk analysis within the Defense program and raise the understanding of the value of energy efficiency investments relative to DoD mission success.

The Deputy Under Secretary of Defense for Logistics and Materiel Readiness (DUSD (L&MR), has the authority to serve as the central administrator of the policies that govern how fuel is supplied to our operational forces, as well as the planning and purchase of all varieties of operational fuel, through the Defense Energy Support Center, a Defense Logistics Agency (DLA) field activity. The DUSD (L&MR) and DLA are beginning to look at the defense logistics implications of biofuels and other synfuels to the force on behalf of the Under Secretary.

The Director, Defense Research and Engineering has responsibility for managing the entire defense science and technology budget and its priorities, including the oversight of Service investments. Significant basic science and applied science work is going into mobile power generation, more efficient materials and engines, light-weight structural materials and alternative operational fuels. While these investments are focused on their potential benefits to DoD operational capability, there is strong historical precedent for DoD science and technology investments advancing the state of the art and sparking commercial innovation.

Finally, the Under Secretary of Defense (AT&L) chairs the DoD Energy Security Task Force. This broadly based group is supported from membership from across the DoD. While this group initially focused on sharing energy-related information within the Department, it's mandate has expanded as its staff and members craft the DoD Energy Security Strategic Plan that has been directed through Deputy Secretary-level guidance for May 2008 completion. Consistent with the Department's own assessments and the DSB Task Force findings, the ESTF is leading and building the proper teams of OSD, Joint staff and Component stakeholders to address DoD processes and practice shortfalls that have created some gaps cited by the GAO. Energy-related technology priorities are also considered by the group and given additional resources based on the ESTF endorsement. Through monthly meetings, the ESTF website and a monthly speaker series, the group has widened the circle of interest in energy issues in DoD and its related interagency, industry and policy communities. The leadership of the ESTF also coordinates its efforts with the wider DoD Installation and Environment community implementing Executive Order 13423, "Strengthening Federal Environmental, Energy, and Transportation Management". At the current state of understanding of DoD's mobility fuel challenges, this outreach approach is adequate for fostering the right discussions and identification of solutions to DoD leaders and staff.

As relevant technology ideas, needed process changes, operational vulnerability concerns or commercial fuel price growth arise from our on-going analysis and interaction, the Office of the Under Secretary of Defense (AT&L) will serve as both a clearinghouse for addressing the issues, and as the Departmental leader in implementing actionable solutions.

RECOMMENDATION 2: The GAO recommends that the Secretary of Defense direct the Deputy Secretary of Defense to establish an overarching organizational framework by directing the executive-level mobility energy official to lead the development and implementation of a comprehensive Department-wide strategic plan for mobility energy. At a minimum, this strategic plan should set forth mobility energy goals and objectives, time frames for implementation, and performance metrics to track and evaluate progress.

DOD RESPONSE: Partially concur. The USD (AT&L), as the DoD executive lead for energy and as supported by the DoD Energy Security Task Force, is overseeing development of a DoD Energy Security Strategic Plan which will be reported out to the Deputy's Advisory Working Group in May of 2008. This plan will provide a DoD-wide strategic construct for considering mobility energy challenges and inefficiencies, as well as for setting goals, objectives, analytic frameworks, actionable metrics and implementation timing for further assessments and execution. The plan is described in context to the management framework described in the DoD Response to Recommendation 1.

RECOMMENDATION 3: The GAO recommends that the Secretary of Defense direct the Deputy Secretary of Defense to establish an overarching organizational framework by ensuring that OSD develop a methodology to enable the full implementation of an energy efficiency key performance parameter in DoD's requirements development process.

DOD RESPONSE: Partially concur. The Department agrees that further clarification of roles and targets may be needed, but the authorities and responsibility to do so already exist under the Under Secretary of Defense (Acquisition, Technology and Logistics) explained in the DoD response to Recommendation 1. Further, the Department will address the pace and required steps for maturing and then implementing the fully burdened cost of fuel analytic methodology currently under development, within the DoD Energy Security Strategic Plan. This foundational analytic work has received significantly more attention over the past six months, and its broader inclusion within the major strategic processes of the Department is indeed a Top 25 Transformational priority for the Department. By taking this methodology through the standard DoD vetting processes, we can ensure the rigor needed to formally implement this is present. After this vetting, it will underpin the Energy Efficiency Key Performance Parameter within the Joint Capabilities Integration and Development System.

RECOMMENDATION 4: The GAO recommends that the Secretary of Defense direct the Deputy Secretary of Defense to establish an overarching organizational framework by ensuring that, as part of its efforts to complete DoD's fully burdened costs of fuel, pilot program, OSD develop an approach for incorporating this cost information into the acquisition decision making process.

DOD RESPONSE: Partially concur. As the report states, efforts are underway to look at the implications of the fully burdened cost of fuel and the logistics implications of delivering fuel for DoD operations in the force planning, requirements generation and acquisition processes. It is ill-advised to focus on the acquisition process alone, as the force planning and requirements processes inform the acquisition process on the value of various qualities desired, which add up to provide, along with well trained personnel, a capability. This capability is provided at an agreed cost and within a certain schedule. Hence, a prioritization towards fuel efficiency must be addressed within certain constraints. Lacking a clearer demand signal, there is little analytical basis for making trades between various qualities, technologies or design options at the acquisition phase. As it is stated in the Department's response to Recommendations 1, a work plan is in development on how best to assess the value of greater fuel efficiency from the operational capability and cost perspectives in DoD equipment and platforms.

That said, significant science and technology investments have also been made, partially directed from the inputs of the Energy Security Task Force, to address the energy efficiency of platforms and hence, operational units.

RECOMMENDATION 5: The GAO recommends that the Secretary of Defense direct the Secretaries of the Army, Navy and Air Force and the Commandant of the Marine Corps to designate an executive-level official within each of their Services to act as a focal point on Department-wide mobility energy efforts as well as provide leadership and accountability over their own efforts.

DOD RESPONSE: Partially concur. The issue of governance and oversight of energy matters within the Military Departments will be raised once the DoD Energy Security Strategic Plan is briefed to DoD senior leaders in May 2008. Until that time, it is inappropriate to pre-judge the deliberative work currently going into that Plan.

REFERENCES

[1] Energy consumed at fixed installations, referred to as facility energy, accounts for most of DOD's remaining energy use.

[2] This concept is known as fully burdened cost, which DOD defines as the total ownership cost of buying, moving, and protecting fuel in systems during combat.

[3] key performance parameter is an attribute or characteristic of a system that is considered critical or essential to the development of an effective military capability.

[4] GAO, Department of Energy: Oil and Natural Gas Research and Development Activities, GAO-08-190R (Washington, D.C.: Nov. 6, 2007).

[5] For a discussion of issues surrounding peak oil production, see GAO, Crude Oil: Uncertainty about Future Oil Supply Makes It Important to Develop a Strategy for Addressing a Peak and Decline in Oil Production, GAO-07-283 (Washington, D.C.: Feb. 28, 2007).

[6] For a more complete overview of U.S. energy challenges, see GAO, 21st Century Challenges: Reexamining the Base of the Federal Government, GAO-05-325SP (Washington, D.C.: February 2005), and Meeting Energy Demand in the 21st Century: Many Challenges and Key Questions, GAO-05-414T (Washington, D.C.: Mar. 16, 2005)

[7] DOD Directive 4140.25, *DOD Management Policy for Energy Commodities and Related Services* (Apr. 12, 2004). Other fuel-related policy documents include DOD Directive 5101.8, *DOD Executive Agent for Bulk Petroleum* (Aug. 11, 2004); DOD Manual 4140.25-M, *DOD Management of Bulk Petroleum Products, Natural Gas and Coal* (Jun. 12, 2002); and Joint Chiefs of Staff, Joint Pub. 4-03, *Joint Bulk Petroleum and Water Doctrine* (May 23, 2003).

[8] Defense Science Board Task Force on Improving Fuel Efficiency of Weapons Platforms, *More Capable Warfighting Through Reduced Fuel Burden* (January 2001).

[9] The JASONs, *Reducing DOD Fossil-Fuel Dependence*, JSR-06-135 (September 2006).

[10] LMI, Transforming the Way DOD Looks at Energy: An Approach to Establishing an Energy Strategy, Report FT602T1 (April 2007).

[11] Defense Science Board Task Force on DOD Energy Strategy, *More Fight—Less Fuel* (February 2008).

[12] Office of the Under Secretary of Defense for Acquisition, Technology, and Logistics Memorandum, "Fully Burdened Cost of Fuel Pilot Program," April 10, 2007.

[13] Joint Chiefs of Staff Instruction 3170.0 1F, Joint Capabilities Integration and Development System (May 1, 2007), and Joint Chiefs of Staff Manual 3170.01C, Operation of the Joint Capabilities Integration and Development System (May 1, 2007).

[14] GAO-07-283.

[15] GAO, Results-Oriented Cultures: Implementation Steps to Assist Mergers and Organizational Transformations, GAO-03-669 (Washington, D.C.: July 2, 2003).

[16] DOD Directive 4140.25, and DOD Directive 5134.01, Under Secretary of Defense for Acquisition, Technology, and Logistics (Dec. 9, 2005).

[17] GAO-03-669.

[18] The 3-Star Group within DOD's Planning, Programming, Budgeting, and Execution process includes members from OSD's Director of Program Analysis and Evaluation; OSD's under secretaries of defense; the Joint Staff Director for Structure, Resources, and Assessment; and the military services' 3-Star programmers. This group addresses major issues and presents decision options to the Secretary of Defense.

[19] GPRA (Pub. L. No. 103-62, (1993)) is the centerpiece of a statutory framework that Congress put in place during the 1990s to help resolve the long-standing management problems that have undermined the federal government's efficiency and effectiveness and to provide greater accountability for results. For additional information, see GAO, *Agency Performance Plans: Examples of Practices That Can Improve Usefulness to Decisionmakers*, GAO/GGD/AIMD-99-69 (Washington, D.C.: Feb. 26, 1999).

[20] GAO, Defense Management: Opportunities Exist to Improve Implementation of DOD's Long-Term Corrosion Strategy, GAO-04-640 (Washington, D.C.: June 23, 2004).

[21] Energy Independence and Security Act of 2007, Pub. L. No. 110-140, § 933 (2007).

[22] Pub. L. No. 109-364, § 360 (2006), states that it shall be DOD's policy to improve fuel efficiency of weapons platforms, consistent with mission requirements, in order to enhance platform performance, reduce the size of fuel logistics systems, reduce the burden high consumption places on agility, reduce operating costs, and dampen the financial impact of volatile oil prices.

[23] GAO, Agencies' Strategic Plans Under GPRA: Key Questions to Facilitate Congressional Review, GAO/GGD-10.1.16 (Washington, D.C.: May 1997).

[24] The *National Military Strategy*, signed by the Chairman of the Joint Chiefs of Staff, is guided by the goals and objectives contained in the present *National Security Strategy* and serves to implement the Secretary of Defense's *National Defense Strategy*. The *Quadrennial Defense Review*, prepared by the Secretary of Defense every 4 years, assesses the nature and magnitude of the political, strategic, and military risks associated with executing the missions called for under the *National Defense Strategy*.

[25] GAO-03-669.
[26] DOD Instruction 5126.47, *Department of Defense Energy Policy Council*, Under Secretary of Defense for Acquisition, Technology, and Logistics (Dec. 2, 1985).

INDEX

A

B

C

D

E

H

I

N

O

P

Q

R

S

T

U

V

W

Y